PLANT NUTRIENT DISORDERS 4
Pastures and Field Crops

PLANT NUTRIENT DISORDERS 4
Pastures and Field Crops

R.G. Weir

G.C. Cresswell

Biological & Chemical Research Institute

NSW Agriculture

Inkata Press

MELBOURNE • SYDNEY

INKATA PRESS
A DIVISION OF BUTTERWORTH-HEINEMANN

AUSTRALIA	BUTTERWORTH-HEINEMANN	271–273 Lane Cove Road, North Ryde 2113
UNITED KINGDOM	BUTTERWORTH-HEINEMANN Ltd	Oxford
USA	BUTTERWORTH-HEINEMANN	Stoneham

National Library of Australia Cataloguing-in-Publication entry

Weir, R.G.
  Pastures and field crops.

ISBN 0 909605 92 0.
ISBN 0 909605 88 2 (series).

1. Nutritionally induced diseases in plants. 2. Pastures. 3. Field crops – Diseases and pests. 4. Field crops – Nutrition. I. Cresswell, G. C. (Geoffrey Charles), 1953– . II. Title. (Series: Plant nutrient disorders; 4).

632.3

Designed by John Van Loon

Typeset in 10½ pt Goudy Old Style by MACKENZIES, Melbourne

Printed through Bookbuilders

CONTENTS

Acknowledgments

The authors wish to thank those people who have contributed photographs to this book: Jeff Hirth, Research Institute Rutherglen Victoria; Dr Peter Hocking, CSIRO Canberra; John Gartrell, Department of Agriculture, Western Australia; and officers of NSW Agriculture – Gordon Stovold (Rydalmere), Tony Dale (Gunnedah), Bob Colton (Orange), Helen Burns (Lockhart), Harry Marcellos (Tamworth) and Brian Dear (Wagga). Thanks are also due to Messrs G.G. Johnson (Orange) and L. Turton (Rydalmere) for duplicating and processing the photographs.

PREFACE

*T*his manual is part of a series called Plant Nutrient Disorders that aims to help farmers, advisers and students to identify nutrient deficiencies and toxicities in a wide range of crops and pasture plants.

In all, there are five manuals, dealing with the identification of nutritional disorders and they are:

1 Temperate and Subtropical Fruit and Nut Crops
2 Tropical Fruit and Nut Crops
3 Vegetable Crops
4 Pastures and Field Crops
5 Ornamental Plants and Shrubs

Each book describes and shows, with the aid of coloured plates, typical symptoms of the common nutrient deficiencies and toxicities found in a range of common crops. The techniques for distinguishing symptoms caused by a nutrient deficiency, toxicity or some other problem, such as a virus disease, herbicide damage or moisture stress, are explained and illustrated.

These methods evolved over thirty years of close collaboration between plant nutrition chemists, advisory horticulturists and agronomists from the New South Wales Department of Agriculture in their efforts to find solutions to many farming problems. This consultancy service ceased in 1988, but it is hoped that the skills developed for diagnosing nutrient disorders in plants will be passed on through each of these books.

Many of the coloured plates used in the manuals are of field specimens collected from affected crops where leaf analysis was used to identify the problem. Analytical records go back to 1958 and cover probably the widest range of crops available to date.

This series provides leaf analysis standards and sampling methods on a wide range of crops, including data for many poorly documented species as well as the more common crops. Where months are mentioned they refer only to the Southern Hemisphere. The corresponding time of season or stage of crop development for the Northern Hemisphere follows in brackets. The booklets also explain how the leaf analysis technique can be used, how to interpret analyses and how leaf analysis should not be used. This information should be helpful to anyone who wishes to build diagnostic and interpretive skills.

CHAPTER 1

PLANT NUTRIENT NEEDS

Sixteen elements are known to be essential for the normal growth of green plants. Three of the elements – carbon, hydrogen and oxygen – are obtained from the air, and from water in the soil. The remainder – nitrogen, phosphorus, potassium, sulphur, calcium, magnesium, iron, manganese, copper, zinc, boron, molybdenum and chlorine – are obtained chiefly from the soil. Other elements, such as silicon and sodium, can improve the growth and health of certain plants but are of little importance for the successful production of field grown crops or pasture plants.

Plants need large amounts of nitrogen, phosphorus, potassium, sulphur, calcium and magnesium which are called 'major' or 'macro' nutrients. The remaining elements are needed in much smaller amounts, and are known as 'minor' or 'trace' elements.

All nutrients are normally present in the soil to some extent. They come from the parent rock, from decomposing organic matter and from fertilisers. Plants absorb these nutrients from the soil solution and transport them to leaves and other parts where they are used for growth and tissue maintenance. Each nutrient is involved in physiological processes essential for plants to function normally. Consequently, if there is too little or too much of any one nutrient, plant health suffers, leading to slow growth, low yields or poor quality, and symptoms are exhibited on leaves or other tissues. There are two types of nutrient disorder:

- **Deficiencies**   where there is too little of an essential element to sustain optimum plant performance.
- **Toxicities**   where supplies of an element are more than the plant can tolerate.

Figure 1 shows the general relationship between the supply of an essential element and crop performance.

Causes of nutrient deficiency

The two main reasons for a deficiency are that a soil nutrient is low or that the nutrient is not in a form available for plant uptake.

- **Very low levels of nutrient in the soil**

Many nutrients are readily leached from the soil by water and over a very long time reserves can become depleted. Cultivation of soils can

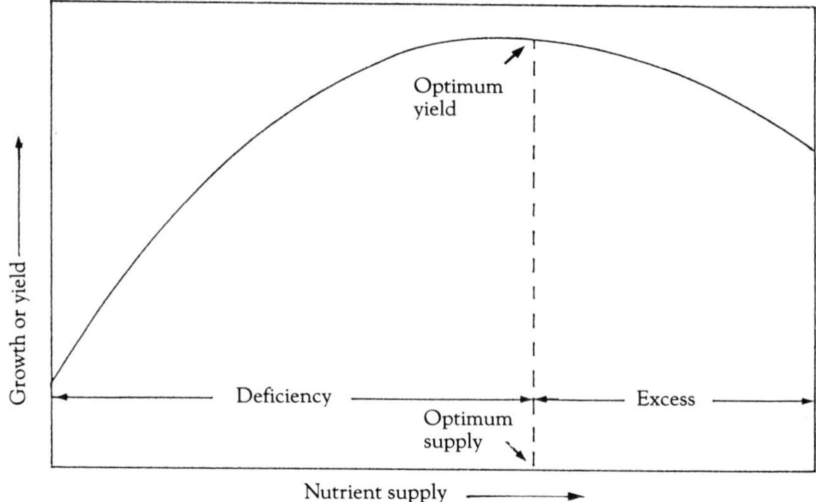

Figure 1   Relationship between crop growth or yield and nutrient supply

accelerate nutrient losses by keeping the ground bare and vulnerable to leaching and erosion. The nutrients removed from the farm in produce or transferred in excrement can also contribute significantly to losses, unless they are regularly replenished by balanced fertilisers and manures. Some soils, especially those derived from sedimentary rocks like sandstones and shales, are naturally poor in nutrients. The most fertile soils are formed from weathered igneous rocks rich in minerals, or from mineral-rich material deposited on river flood plains.

Many soils are poor because of their sedimentary origins and their long history of leaching. Deficiencies of nitrogen, phosphorus, sulphur, magnesium and boron are widespread in these soils.

- **Insufficient level of available nutrient in the soil**

Plants absorb nutrients primarily from the soil solution so nutrient forms with low solubility, for example some mineral phosphates, are not readily available to the plant. The acidity or alkalinity of the soil (measured by its pH), affects nutrient availability. Deficiencies are more likely to develop in soils that are alkaline or very acid than in soils where the pH is neutral or only slightly acid. For example, zinc and iron deficiencies occur in sensitive crops planted in many of the naturally alkaline soils in inland New South Wales, while molybdenum deficiency is often found in crops and pasture legumes grown on coastal acid soils.

Soil pH can be altered by management, for example, deficiencies of boron, iron, manganese and zinc can develop in soils made too alkaline by over-liming. If soils become too acidic, the availability of nutrients like calcium, magnesium, nitrogen, phosphorus and molybdenum is reduced. Excessive acidity, a consequence of the long use of sulphate of ammonia, for example, will reduce the availability and lead to deficiencies of these nutrients.

Causes of nutrient toxicity

Salinity and soil acidity are the two major causes of toxicities in crops and pastures. Under saline conditions, chloride and sodium toxicities are common. In acid soils, toxicities of manganese and aluminium often occur.

Many soils have naturally high levels of salt, especially in arid and semi-arid areas, where rainfall is too low to leach it away. Salinity problems quickly develop in these areas if the native timber is cleared or irrigation is used without adequate drainage. These practices often lead to a rising watertable which brings salts up into the root zone. Salts of chloride and sodium leached from soil by irrigation water will be concentrated by evaporation along edges of furrows, in low areas or in seepage areas above impervious subsoil layers.

Some salt- or boron-affected soils have developed from rocks formed under seas or large inland lakes which became saline as they dried up. Irrigation water from bores, wells, dams or rivers, especially at times of low river flow, are other sources of chloride and sodium. Any water to be used for irrigation should be tested for salinity. Although fertilisers are not a major factor in salinity, avoid muriate of potash (potassium chloride) and fertilisers containing it where salinity is a concern, as they may add to the salinity problem.

Many soils are naturally acid and may become even more acidic after several years of cultivation, irrigation and fertiliser use. This can lead to an increased availability of elements such as manganese and aluminium which may cause toxicity problems. High manganese availability can also be a problem in waterlogged soils and in some soils affected by heat and drying during summer fallow. The solubility of aluminium increases in strongly acid soils causing severe stunting of the roots of sensitive plants such as lucerne and barley.

Nitrogen is one of the most important of the nutrients normally applied to crops. However, most nitrogenous fertilisers – ammonium sulphate, urea, ammonium nitrate and mixed fertilisers containing ammonium salts – acidify the soil. Ammonium sulphate (sulphate of ammonia) is the most acidifying and should be avoided for most soils. Even the nitrogen fixed by legumes, or from manures can contribute to soil acidification. This happens because some of the plant-available nitrogen is leached from the soil as soluble nitrate, a process which also removes calcium and magnesium.

Toxic concentrations of salts accumulating near the roots from careless or excessive use of fertilisers can also cause injury. Germinating seedlings and young plants are especially sensitive when concentrated soluble fertilisers are banded too close to the roots.

Under special circumstances, toxicities of some of the trace elements can occur. Boron, copper and zinc have a narrow optimum range, between enough and too much, and should be used carefully. Sawdust derived from softwood timber treated with boron for pest control has caused toxicity when used as mulch around sensitive plants. Heavy metal residues in mining and sewage wastes are other potential causes of toxicity.

Uneven application or use of high rates of trace elements can easily lead to toxicity effects. When applying borax, for example, all lumps should be crushed and the powder spread evenly over the area to be treated to avoid pockets of high boron concentration. Foliar sprays of trace elements can cause injury if product recommendations are not followed. Soluble copper salts, for example, can be very phytotoxic when applied as sprays and are often 'neutralised' with hydrated lime to avoid foliage burn.

Nutrient imbalances

Plants need a proper balance between the various nutrients to grow well. For example, too much potassium may interfere with the uptake of magnesium or calcium, leading in some cases to a deficiency of one of these elements. In the same way, an excess of phosphorus fertiliser can induce zinc or iron deficiency if these elements are in short supply.

Crop sensitivity and tolerance

Plants differ in their sensitivity to specific nutritional problems. For example, boron deficiency is more common in sunflower, canola (rapeseed) and lucerne than in wheat. Zinc deficiency is more likely to affect maize, linseed or soybeans than sunflowers, while lucerne, canola and soybean are more sensitive than oats or phalaris to high levels of manganese in the soil.

Differences in plant sensitivity to a deficiency arise because some species or varieties need more of a nutrient for growth. Others are less able to absorb enough of an element from the soil. Crops also differ in their tolerance of toxicities. This is related to differences in their abilities to exclude elements from uptake at the root surface, to keep absorbed elements away from sensitive tissues, or to render potentially toxic elements less biologically harmful.

CHAPTER 2

IDENTIFYING NUTRITIONAL PROBLEMS

*T*he visual symptoms plants exhibit in response to nutrient deficiency or toxicity are a useful guide for identifying the cause of a disorder. Common plant responses include unusual colours or patterns in the leaves, burns, distortion of individual plant parts, stunting or abnormal growth.

Table 1   Major causes of visual symptoms in plants

Nutritional disorders	Other disorders
deficiencies	infectious diseases – fungal, bacterial or viral
toxicities	insect damage
	physiological–environmental stresses
	mechanical injury
	chemical injury – pesticide, air pollution,
	spray burn

As some non-nutritional factors can also produce similar symptoms (Table 1), careful observations are needed to ensure the diagnosis is reliable.

In higher plants, nutrients move from the roots to other parts of the plant through a network of cells called the vascular system. These tissues are specialised in the transport of water, nutrients and metabolic products throughout the plant. The most obvious part of this vascular system are the veins in leaves. The arrangement of veins and the ease with which individual elements move within the plant (mobility) have a strong influence over the way symptoms develop. In manganese-deficient leaves, for example, leaf blade tissues close to the major veins are last to become chlorotic because they have first call on manganese available in the sap.

The close relationship between the symptom pattern and the arrangement of the veins is an important feature of nutrient-related problems which distinguishes them from most symptoms of non-nutritional problems. The latter usually show no relationship to vein pattern.

Characteristics of nutritional symptoms on leaves

- Restricted initially to a single leaf-age class, that is, young, old or intermediate aged leaves.
- Patterns are symmetrical and closely related to leaf venation.
- Changes in leaf colour and tissue death develop gradually (rarely overnight).
- Boundaries between green and chlorotic areas on a symptom leaf tend to be diffuse. Strong, definite boundaries are often produced by herbicides or viruses.
- Leaf patterns are rarely blocky or angular. Such patterns can be caused by a pathogen or occasionally by nematodes.
- Damage to the surface of a symptom leaf is unusual. Nutritional problems impair cell function and rarely cause mechanical disruption of the cuticle.
- Symptoms develop first in tissues most distant from the major veins of the leaf, for example, the interveinal regions, tips and margins of the leaf blade.

After establishing that the symptoms are nutritional, take the following steps to identify the disorder and its underlying cause.

STEP **1** –

Gathering the facts

Background information is needed to define the problem and to identify the most likely factors contributing to its development. Some causes can then be discounted, narrowing the scope of the investigation. Furthermore, this information can help establish the underlying cause of a problem, which is crucial when developing an appropriate corrective treatment.

Field description   Briefly describe the overall appearance of the crop, highlighting obvious abnormalities such as unusual leaf colouring, stunting, or thinning in the crop. Relationships between the distribution of problem areas in a crop and geographical features, such as soil type, fence lines, crop rows or irrigation bays, may point to a soil or nutritional cause.

Severity of problem   Estimate the percentage of the crop currently affected and the expected reduction in yield. This information may be useful for deciding between possible causes. For example, a mild magnesium deficiency is unlikely to be the primary cause if yield has been reduced by 50%.

Crop   Details of cultivar and overall crop health are needed for interpreting plant test results. Sensitivity to a particular nutrient deficiency or toxicity often differs with cultivar.

Crop developmental stage   For example, vegetative, flowering, seed development or ripening. Knowledge of the plant maturity at sampling is essential information to allow an accurate interpretation to be made of plant test results and for formulating sensible corrective measures.

Contributing factors   Any change in management immediately before the symptoms developed should be noted. The time taken for the symptom to become fully expressed may also provide a clue to the cause. Symptoms which develop suddenly, for example, are often caused by catastrophic events such as frost, wind, pests, and herbicide or other chemical sprays.

Cropping history   Previous crops grown on the site, their management and performance. Residual chemicals in the soil, such as herbicides, liming materials and fertilisers, can affect later crops.

Soil type and depth   Are the soils uniform in type and depth across the problem area? Is there any relationship between the distribution of affected plants and variation in soil properties? Are affected patches related to land levelling, former roadways, fence lines, irrigation banks, or log burning. Relationships between the occurrence of a disorder and soil type or previous land use often indicate soil chemical or physical problems. Dig down and examine the soil at plough depth for a possible hard pan; compare the soil below poor patches of crop with that in better grown areas.

Irrigation type/frequency   Is irrigation practised and, if so, what type: overhead, flood, trickle or capillary? Has the crop been moisture stressed? Severe water stress can trigger early senescence in a crop which may be confused with a nutritional problem.

Drainage   Are the soils well drained? Waterlogging can lead to nutritional disorders including manganese toxicity, iron deficiency and nitrogen deficiency, or produce wilting, leaf fall, vein chlorosis and hormone-like symptoms of new growth.

Weather conditions   Were weather conditions leading up to the problem unusual, for example, periods of heavy rain, drought, high or low temperatures, or frost? Both current weather and that occurring some months earlier may be significant.

Fertiliser history   What fertilisers have been used for this and previous crops? Problems can sometimes be traced to a recent change in fertiliser practice, to a history of over- or under-use of fertiliser, or of an unbalanced program. The current program should be compared with programs used by other growers in the area.

Paddock history   Does the problem area coincide with a previously cropped, fertilised or flooded section of the block?

Spray program   What chemical sprays have been used on this crop – pesticides, nutrients or other? Pesticide and nutrient sprays can burn the foliage and affect flowering, if not used according to the manufacturer's recommendation. Nutrient sprays, and some fungicides that contain trace elements, can contaminate samples collected for leaf analysis and confuse the diagnosis.

Plant health   Are plants diseased or infested with insect pests?

STEP **2** –

Diagnosis from visible symptoms

Visible changes in a crop, such as yellowing, small leaves and poor seed set, all begin as a breakdown in cell functioning and tell us that a nutritional disorder exists. The distortion of new tissues or flowers, or the death of growing points, typical of boron deficiency, for example, occurs because boron is necessary for the proper regulation of cell division. Similarly, the leaves of nitrogen- or magnesium-deficient plants are pale because nitrogen and magnesium are constituents of chlorophyll. Such links, between an element's physiological function and a specific abnormality which results when it is deficient, are common in plants. For this reason the nature of the symptom can provide a useful guide to the identity of a nutritional disorder even in unfamiliar crops.

The two most important diagnostic features of a nutritional symptom are where the symptom is found on the plant (location) and its appearance (colour and pattern).

Description of symptoms   After recording your field description of the problem in the crop, you must now examine symptoms on individual plants.

Location   Where do the symptoms first appear on the plant, for example, young, mature or old leaves, or on the stem, roots or seed? Nutritional symptoms generally do not develop uniformly over a plant but show first in specific organs such as the leaves, fruit, roots, shoot or growing point. Leaf symptoms, the most widely used diagnostic feature, can occur in the upper, middle or lower sections of a plant, depending on the mobility of the element. Mobile elements like nitrogen, magnesium or potassium are moved about the plant relatively easily to satisfy local shortages, particularly in new shoots or developing seeds. When one of these mobile elements is deficient, the **older leaves** are the first to be depleted and first to show symptoms.

Less mobile elements, like iron, boron or calcium, do not move readily from old to younger tissues, so when they are deficient the symptoms appear in the **newer or upper leaves,** the flowers or seed. Even a temporary shortage of one of these immobile nutrients causes the young tissues growing at the time to suffer and develop symptoms.

Symptoms of nutrient toxicity generally show first in the oldest leaves. These leaves have the highest transpiration rates and receive most of the nutrients absorbed by roots as they move in the transpiration stream.

Pattern   Study the development and appearance of the symptoms. Observe the size and shape of the plant, the overall foliage colour, the colour of symptom leaves and the pattern of chlorotic (pale) or necrotic (burnt) areas in relation to vein pattern. Also note any irregular shape, splitting, cracking or corkiness of affected organs. All of these may help to establish the identity of the disorder.

Deficiency or Toxicity?   The first question to answer is: Do the symptoms indicate a deficiency or a toxicity? The following generalisations may help to answer this question.
- Deficiency symptoms typically occur on a single leaf-age class unless, of course, more than one problem exists.
- Toxicities commonly produce burnt or necrotic (dead) areas of tissue. When this occurs on new leaves it almost always indicates a toxicity. Burns on old leaves may or may not be due to toxicity, for example, potassium and magnesium deficiencies produce necrosis on older leaves in many species.
- Toxicity symptoms often develop rapidly. When this happens, the affected leaf tissue may change from healthy green to grey-green or dark brown without a transitional yellow phase.

Symptoms on both old and new leaves may indicate a toxicity. When an excess of one element causes a nutrient imbalance, deficiency symptoms may be seen in the young leaves while older leaves may show burn or other symptoms of toxicity. Excess phosphorus, manganese or zinc can induce the chlorosis of iron deficiency in young leaves as well as symptoms of nutrient excess in the old leaves.

Which deficiency?   If a deficiency is suspected, the location of symptoms on the plant is a useful guide to whether the responsible element is mobile or immobile. The way symptoms develop on leaves, shoots, flowers or fruit provides further clues. Small, irregularly shaped leaves, shortened internodes, aborted flowers, poor seed set and distorted fruit can be characteristic of a particular deficiency.

Diagnostic keys, similar to the one in Table 2, provide a framework for a visual diagnosis.

Although visual symptoms are useful for diagnosis there are two major weaknesses:
- Clear visual symptoms do not usually appear until a disorder is quite advanced and some loss of yield or quality has occurred. By this stage, even prompt remedial action will not reverse losses. Also, the absence of symptoms in a crop or pasture does not mean that nutrition is adequate. 'Hidden hunger' is the condition where performance is limited but no symptoms have been expressed.
- Visual symptoms can be unreliable when more than one element

is limiting or when some environmental stress has modified the normal pattern.

Table 2   Quick guide to nutrient deficiencies – what to look for

---

Symptoms first seen in *older* leaves

Leaf coloration even over whole leaf

Nitrogen	Pale green to yellow leaves.
Phosphorus	Leaves dull, lacking lustre, bluish green or purple colours. Poor growth.

Leaf coloration forms a definite pattern

Potassium	Scorching and yellowing, commonly around the edges of leaves, which may become cupped.
Magnesium	Patchy yellowing often with a triangle of green remaining at the leaf base. Sometimes brilliant red or orange patterns or scorching.

Symptoms first seen in *young* leaves

Leaf coloration forms a pattern

Iron	Almost total loss of green between veins, leaving faint green 'skeleton' of veins on leaf.
Zinc	Severe restriction of leaf size or stem length, or both (hence the terms 'little leaf' or 'rosetting'). Distinct interveinal creamy yellow patches on leaves in many species.
Copper	Tip leaves cupped, narrow, distorted or scorched. Defoliation from tip. Chlorosis interveinal or irregular.

Symptoms first seen in either *old* or *young* leaves

Leaf coloration forms a pattern

Manganese	Mottled diffuse pale green to yellow patches between veins. No restriction of leaf size (unlike zinc).

Symptoms usually most prominent in other tissues – seen first in youngest tissues and fruit

Calcium	Breakdown of parts of fruit in some species. Collapse of flower stalk (flax, rapeseed) or leaf petiole (clover).
Boron	Internal cracking or breakdown of root or stem tissues or flower stalk. Irregular shaped tissues, corkiness or surface cracking of stems. Irregular flower development or poor seed set.

STEP **3** –

Confirming the diagnosis

Where there is uncertainty about the visual diagnosis, or where an incorrect diagnosis would prove costly, other techniques can be used to confirm the diagnosis.

Trial treatment of a portion of the crop   This is the most direct method of confirming a diagnosis. Providing some part of the crop is left untreated to gauge treatment effectiveness this method gives the most certain answer. But trials can be slow and, if unsuccessful, provide no new information which could lead on to a correct (successful) diagnosis.

Plant tissue analysis   Tissue analysis is a powerful diagnostic tool which provides good direct evidence of the nutritional status of a crop. It can be used to verify a visual diagnosis and, because it tests for other nutrients, it enables a new interpretation if the original symptom-based diagnosis is wrong. However, plant analysis may not reveal the underlying cause of a disorder.

Plant analysis is used to

- Diagnose nutrient deficiencies and toxicities:
    confirm a diagnosis based on symptoms,
    identify 'hidden hunger', and
    suggest additional tests to identify a problem.
- Predict nutrient disorders in current or future crops.
- Develop and adjust fertiliser programs.
- Measure the amounts of nutrients removed in crop
  produce and residues with the view to replacement.
- Survey the nutrient status of a crop throughout a district.
- Compare the nutritional status of soils or growing media.
- Estimate the dietary value of a crop or pasture.

Soil and water analysis   These can help determine the cause of crop nutrient problems, but should not be used alone to identify the disorder. Soil analysis is the only means of forecasting crop nutrient needs before planting, providing the soil test is calibrated to both the crop and soil type. For example, a soil test calibrated for wheat or for pasture could be misleading if used for a maize or lupin crop. Likewise, quick sap tests of leaves or other tissues, which have not been correctly calibrated for a particular crop or pasture species, will not give a reliable guide to its nutrient status.

As no one diagnostic procedure is entirely satisfactory, the best approach is to use a number of techniques to develop a diagnosis which is supported by both laboratory tests and field observations.

Correcting the problem

Corrective treatments should always aim to solve the underlying cause of a problem and not simply reduce the immediate effects on the crop. For example, potassium would not be very effective in correcting low potassium uptake caused by high sodium in the soil, unless steps were taken to correct the salinity problem. **Unless the true cause of a crop disorder is correctly established, little will be**

gained from the experience, and you will be no better able to deal with the problem when it is next encountered.

step **5 –**

Following up

The only sure way of knowing whether a correct diagnosis has been made and of improving diagnostic skills is to see whether a crop has responded to a corrective treatment.

CHAPTER 3

PLANT ANALYSIS

Plant analysis involves chemical testing of plant tissues (usually leaves) to estimate how effectively elements have been taken up by the plant. It gives the current nutrient status of a crop but should not be used to forecast future supplies to the crop. It, therefore, differs from soil analysis which aims to predict the ability of the soil to supply nutrients.

The plant analysis method of determining the nutrient status of a crop compares the concentration of each element in the tissue sample with predetermined ranges (leaf standards) for healthy, productive crops of the same species (see Appendix A). Critical levels are the criteria used to decide whether each element is adequately supplied to a crop. It is the leaf level found when shortage of a nutrient just begins to limit growth (or quality). The term is usually defined as 'the tissue concentration associated with a 10% reduction in yield caused by a slight deficiency of an element' (Figure 2). As plant performance can be reduced by toxicity as well as by deficiency, both upper critical (toxicity) levels and lower critical (deficiency) levels can be established for most nutrients.

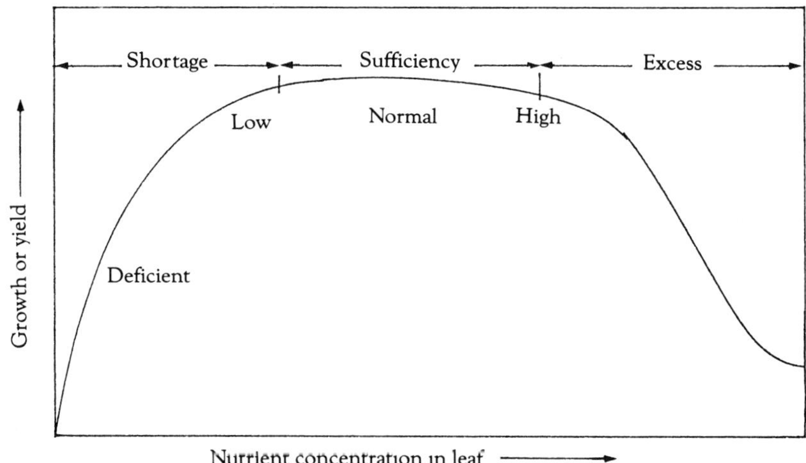

Figure 2   Relationship between plant growth or yield and leaf nutrient concentrations

Leaf standards can also be given in terms of a set of ranges which classify leaf levels as deficient, low, normal, high or toxic. Such ranges better describe the nutritional status from deficiency through normality to toxicity than is achieved with a single critical value. They can also indicate the likelihood of symptoms or quality loss occurring in the crop. High quality is often just as important as top yields and rapid growth for successful crop or pasture production. The five classes of plant nutrient status normally defined are:

- **Deficient**   Symptoms present – nutrient is too low for optimum performance.
- **Low**   No symptoms but nutrient may be too low for optimum crop performance and quality.
- **Normal**   No symptoms – level is adequate.
- **High**   No symptoms – level higher than necessary which may cause imbalance or loss of quality.
- **Excess**   Level too high for optimum performance – toxicity symptoms may be present.

The nutrient composition of a leaf changes during the season; it will vary with its age and its position on the shoot as well as the physiological stage of the plant, for example, vegetative, flowering, fruiting or tuber development. Taking the correct type of leaf at the right stage of the crop's growth is essential; **leaves must be sampled correctly or the analytical results will be misleading.**

Sampling method

General directions for most crops are given below. Detailed procedures for individual crops are given with the nutrient standards in Appendices A and B.

Tissue   Use whole leaves, that is, blade (lamina) including midrib, but often removing or shortening extended petioles. Sample only clean healthy plants of normal growth characteristics.

Leaf age   A fully expanded, recently matured leaf.

Number of plants   Take leaves from about 25 plants throughout a uniform and representative section of a block within one soil type, following a zigzag pattern throughout the sampled area.

Number of leaves per sample   60 to 100

Leaf analysis data for some other crops

For many crops there are no published standards or the standards are incomplete. Appendix B gives tentative guidelines for some of these crops based on diagnostic and research analyses conducted at the New South Wales Department of Agriculture laboratories for over 30 years.

The 'Normal' range shows analyses from records for healthy productive crops with no symptoms. The 'Deficient' range refers to analyses recorded for crops with deficiency symptoms (or otherwise established to have been deficient), including analyses from research trials. This range is probably not completely defined by the data listed. The 'Excess or toxic' range includes data from crops either showing toxic symptoms or where a serious nutritional imbalance was suspected.

In Appendix B, 'Below normal' and 'Above normal' replace the 'Low' and 'High' ranges which were used in the earlier set of tables (Appendix A), for which more complete information was available. Although these values are not considered 'Normal', loss of yield or quality due to nutrition has not been definitely established.

Making a two-sample comparison

The rapid growth rate, quick maturing, short-lived nature of annual crops combined with the day-to-day effects of cultural treatments, such as irrigations or fertiliser side dressings, and the effects of grazing on pasture plants makes an accurate assessment of subclinical deficiencies or toxicities more difficult than it is for perennial crops such as citrus, apples or grapes. The effects of these factors can be reduced by comparing the main sample with a second sample taken from a better growing patch within the crop. Alternatively, to identify the cause of poor growth in a bad patch within a crop, take separate samples of plants of the same age from both the healthy and affected areas. Where no leaf standards exist, or their use is inappropriate because sampling could not take place at the correct time or in the specified manner, this type of comparison of nutrient levels in leaves of healthy and affected plants will help in diagnosing a nutritional disorder (see Table 3).

Quick sap tests

Because of the short duration of many crops and the need for a quick answer to benefit the current crop, many workers have examined the effectiveness of measuring the soluble fraction of nutrients in the sap, rather than the total amount in the whole of the leaf tissue. This approach uses the current flow of a nutrient into tissues as a guide to supply. The soluble nutrient can be estimated either in the laboratory, or more roughly in the paddock, from the colour intensity of a tube of solution or a chemically impregnated strip of paper.

Merckoquant test strips (from E. Merck of Germany) are one of the simplest tests to use. Sap is squeezed from the petiole of a selected leaf onto a square of paper impregnated with chemicals sensitive to the particular nutrient being tested (usually nitrate or sometimes potassium). After a short period for the chemical reaction to occur, the concentration of nutrient in the sap is determined from the colour intensity of the strip compared with standard ranges for the nutrient expected in healthy crops.

Because sap tests measure nutrients present in conducting tissue, such as a leaf petiole or portion of stem, they can provide an index which reflects the current nutrient supply. However, the tests are fraught with difficulty. Apart from the problems of accurately controlling the testing conditions in the paddock, even greater error may come from the selection of the sample tissue and from variability within the crop. The flow of soluble nutrients within the conducting tissue is extremely changeable, being sensitive to many factors besides the nutrient supply, such as the time of day when the sample is taken, shading of the sampled leaf, an overcast sky, or the time since the last irrigation. Soluble nutrients quickly rise and then fall after a recent fertiliser side dressing and can be influenced by drought or another nutrient disorder. The quick tissue test usually measures only a single element and can be misleading if another nutrient is deficient or toxic.

CHAPTER 4

COMMON NUTRITIONAL PROBLEMS AND THEIR CORRECTION

Many crops, because of their rapid growth and short growing season, have a high demand for nutrients. Pasture legumes and grasses during periods of high growth and grazing pressure also require a speedy replenishment of nutrients removed by grazing or cutting for silage or hay. Deficiencies of the major elements, nitrogen and phosphorus, are commonly encountered in most districts and crops, where too little fertiliser is used. Severe phosphorus deficiency is common in newly established pastures and new cropping soils. Nitrogen is also a frequent problem in new pastures where legumes have not been nodulated with an effective strain of the nitrogen-fixing bacteria Rhizobium, or where molybdenum deficiency has prevented efficient fixation of nitrogen from the air. Potassium deficiency often develops through its removal in hay cuts or in crop produce where it is not replenished in fertilisers. Sulphur deficiency more often affects pastures than crops, particularly in tableland areas remote from the sea and industrial areas. Other common problems are deficiencies of molybdenum, boron and zinc, and toxicities of manganese, aluminium, chloride and sodium.

Soil acidity is an increasing problem in many areas which can lead to toxicities of manganese and aluminium, or deficiencies of molybdenum, calcium and magnesium. A lack of balance between the nutrient elements or over-fertilisation, particularly the excessive use of nitrogen or potassium, are other causes of nutritional problems in crops and pastures, and the animals which graze them.

Table 3 lists some common nutritional problems, the soils in which they are likely to occur, the crops and pasture plants most often affected, and the levels of nutrients found in the leaves of affected and normal plants.

Table 3   Some common deficiencies and toxicities affecting pasture and crop plants in New South Wales

Crop	Problem	Occurrence in NSW	Symptom	Nutrient level in leaf Affected plant	Normal plant
Lucerne	Phosphorus deficiency	Most districts	Growth stunted, leaves small and dark green with purplish coloration on undersides.	0.14% P	0.29% P
	Potassium deficiency	Many areas, especially lighter soils	Small white spots, yellowing or burn around outer parts of leaflets. Lower leaves affected first and show more acute symptoms.	0.4% K	2.5% K
	Boron deficiency	Tablelands and coast	Upper leaves turn yellow, often with reddish purple tinges. Upper stem stunted giving umbrella-like appearance.	16 ppm B	35 ppm B
	Sulphur deficiency	Tablelands and North Western Slopes	Leaves overall pale green to yellow. Seen first in younger leaves.	0.13% S	0.35% S
	Molybdenum deficiency	Coast and tablelands	Plants stunted with pale green leaves. Lower leaves may die and fall.	0.3 ppm Mo 2.4% N	1.5 ppm Mo 4.3% N
	Chloride excess	Coast and inland	Poor growth, straw-coloured leaf scorch from margin, and tip of leaf extending back towards leaf base. Leaf fall.	2.6% Cl	0.5% Cl
	Manganese excess	Southern Tablelands	Yellowing and death of leaflet margins starting at the tip, with some yellow-grey necrotic spots mostly towards the apex of the leaflet.	1800 ppm Mn	60 ppm Mn
Sub-terranean clover	Nitrogen deficiency	Most districts in new sowings, acid soils and faulty inoculation methods	Pale leaves, stunted growth. Roots lack nodules or have many small whitish nodules.	1.6% N	3.6% N
	Phosphorus deficiency	Most districts, new sowings with history of low fertiliser use, acid P-fixing soils	Very stunted growth, small dark green leaves.	0.15% P	0.29% P

	Sulphur deficiency	Tablelands, especially Northern Tablelands	Pale leaves and poor growth.	0.12% S	0.28% S
	Molybdenum deficiency	Coast, tablelands and slopes	Pale leaves and poor growth.	0.13 ppm Mo 1.6% N	0.60 ppm Mo 3.6% N
	Boron deficiency	Tablelands, slopes and coast	Orange-red leaves and petioles, petioles shortened.	10 ppm B	31 ppm B
	Chloride toxicity	Coast, Hunter and Windsor	Burn of margins on older leaves.	3.2% Cl	0.5% Cl
	Sodium toxicity	Coast, Hunter and Western Plains	Leaf burn.	1.4% Na	0.3% Na
White clover	Potassium deficiency	Coast and tablelands, intensively cropped soils	Yellowing and burn of leaf margin.	0.4% K	1.8% K
	Deficiencies N, P, S, Mo and toxicities Cl, Na	As for subterranean clover			
Wheat and oats	Nitrogen deficiency	All districts, especially light soils previously cropped and low fertiliser N	Stunted with few tillers. Upper leaves pale, lower leaves turning yellow and dying from the tips. Early maturity.	1.3% N	3.6% N
	Phosphorus deficiency	Most districts where current and past P fertiliser usage low	Stunted erect growth. Young leaves dull dark green but older leaves yellow from the tip and die.	0.12% P	0.31% P
	Magnesium deficiency (soil acidity)	Central and Southern Slopes on acid light soils	Mottled chlorosis of leaves. Chlorosis often 'beaded' and leaves die from tips. Patchy yellow crop.	0.06% Mg	0.22% Mg
	Chloride toxicity	Central and North Western Plains, inland irrigation areas	Stunting, yellowing and death of lower leaves. Premature wilting.	3.2% Cl	0.4% Cl
Maize	Nitrogen deficiency	All districts, especially light heavily cropped soils with insufficient fertiliser, waterlogged patches	Poor growth, pale leaves. Old leaves become yellow and burn, beginning at and spreading back from the tip along the centre of the leaf.	1.4% N	3.2% N

continued overleaf

Table 3 (continued)

Crop	Problem	Occurrence in NSW	Symptom	Nutrient level in leaf Affected plant	Normal plant
	Phosphorus deficiency	Mostly coastal areas	Poor growth, purple leaves (most varieties), yellowing from the ends of older leaves.	0.13% P	0.28% P
	Potassium deficiency	Coast and Northern Tablelands	Burning of leaf edge from the tips of older leaves. Stunted with short thin stems.	0.61% K	3.5% K
	Zinc deficiency	North Western Slopes and Plains; calcareous, high pH, high-phosphate soils	Broad yellow bands along the lower half of leaves. Later, interveinal striping and bronzing.	12 ppm Zn	25 ppm Zn
	Molybdenum deficiency	Acid coastal soils where crop sown with low-molybdenum seed	Very pale, stunted seedlings with burnt leaf margins.	Sown with seed <0.08 ppm Mo	Sown with seed >0.08 ppm Mo
Sunflower	Nitrogen deficiency	Repeatedly cropped light soils without prior clover pasture or sufficient fertiliser	Pale small leaves, poor growth. Older leaves turn yellow and senesce.	2.1% N	4.5% N
	Phosphorus deficiency	All low phosphate soils given insufficient fertiliser P	Dark grey necrotic lesions on older leaves.	0.08% P	0.36% P
	Boron deficiency	Tablelands	Stem shortened near top of plant especially near flowerhead. Necrosis of flower stalk and bases of upper leaves.	12 ppm B	28 ppm B
	Molybdenum deficiency	Acid soils of tablelands and slopes	Leaves pale, marginal burn and cupping appearing soon after emergence.	0.04 ppm Mo 12 000 ppm nitrate-N	0.45 ppm Mo 1900 ppm nitrate-N
Soybean	Nitrogen deficiency (nodulation)	Low fertile soils, seed not inoculated before sowing	Stunted growth. Small pale green to yellow leaves which hang downward and die prematurely.	1.3% N	3.1% N
	Phosphorus deficiency	Low phosphate soils given insufficient fertiliser P	Small dull blue-green leaves. Brown interveinal necrotic spots in older leaves.	0.14% P	0.28% P

	Potassium deficiency	Sandy coastal soils subjected to previous heavy cropping	Yellowing and light brown burn around the edge of older leaves.	0.80% K	2.5% K
	Zinc deficiency	High pH clay cracking soils of the North Western Plains	Stunted yellow-brown sections of crop. Leaves small with interveinal yellowing becoming bronzed.	14 ppm Zn (0.3% P)	43 ppm Zn (1.0% P)
Canola (rapeseed)	Nitrogen deficiency	Most soils where inadequate fertiliser N is used, unless following legume pasture	Poor growth, pale green leaves developing red to purple colours then yellow and senescing.	1.2% N	3.8% N
	Phosphorus deficiency	Most districts – low phosphate soils given insufficient P fertiliser	Very stunted growth. Leaves dull grey-green becoming a dull purple colour with age.	0.11% P	0.33% P
	Boron deficiency	Tablelands and Southern Slopes especially light textured acid soils, recently limed soils	Mottled chlorosis and downward curling of leaves. Lower flower and pod numbers.	13 ppm B	25 ppm B
	Manganese toxicity	Acid soils, Southern Slopes wheatbelt	Stunting. Yellow leaf margins, cupping.	3550 ppm Mn	91 ppm Mn
	Aluminium toxicity	Acid soils, low in organic matter	Stunted patches in the crop. Poorly developed stubby roots.	–	–

NITROGEN

Nitrogen is required in large amounts relative to the other essential elements and most cropped soils need regular additions in the form of fertilisers or manures. A nitrogen shortage reduces all aspects of crop and pasture production, namely, growth, yield and quality. For example, in a nitrogen-deficient cereal crop, growth will be stunted, tillering reduced, less grain will set and in smaller heads, grain protein is lowered, and the baking quality of the flour diminished. Likewise, in a pasture, both the yield and the quality of the forage will be reduced if nitrogen is limiting.

Function Nitrogen is an essential constituent of protein and chlorophyll, the green pigment in leaves. It is quite mobile within the plant and, when it is in short supply, will move from older leaves to satisfy the demand of younger leaves, developing grain and new growth.

Deficiency symptoms Too little nitrogen results in uniformly pale leaves. The older leaves, in particular, may yellow and die prematurely. In some crops, such as canola and clover, red or purplish colours develop in paler than normal leaves, while in many grain crops, including sorghum and wheat, the stems and leaf sheaths become red. Growth of the whole plant is reduced and tops are less leafy than normal. Grain or seed production is depressed.

Too much nitrogen can promote excessive vegetative growth and delay seed formation, leading to a loss of grain yield. A strong growth of leaves is often made at the expense of roots in root crops such as sugar beet. The sugar content is also lowered.

Corrective measures Symptoms of nitrogen deficiency may appear when too little fertiliser is applied as basal and side dressings to the crop, or where marked nitrogen losses have occurred. Nitrogen is easily leached from the soil as nitrate, particularly in light sandy soils low in organic matter. Gaseous losses of nitrogen also occur and these are encouraged by waterlogging and surface application of fertiliser. Total application rates of nitrogen fertiliser vary greatly (Table 4), from less than 50 kg N/ha to more than 100 kg N/ha depending on the crop, soil, rainfall, irrigation, plant spacing, cropping level, and previous soil management including legume based pasture or length of fallow. Even leguminous crops, such as field peas, are sometimes given some fertiliser nitrogen to promote rapid early growth. Nitrogen fertilisers commonly used in crop production are listed in Table 5.

Table 4   Rates of fertiliser N, P and K commonly used for some crops (kg/ha)[1]

Crop	Dryland or Irrigated	Basal application N	P[2]	K	Top-dressing N
Pasture	D	–[3]	10–30	60[4]	
(legume-based)	I	–[3]	20–50	50–100[4]	
Pasture grass	D	10–20	5–20	20–40[4]	30–50
	I	20–30	10–40	25–50[4]	30–100
Lucerne	D	0 or 5–10[3]	15–20	60–200	
	I	0 or 5–10[3]	20–50	150–300[4]	
Maize	D	75–150[5]	10–20	30–60[4]	20–40
	I	100–250	30–50	50–100[4]	50
Wheat and barley	D	10–30[5]	10–20	30[3]	
	I	25–50	10–30	30–60[1]	25
Oats	D	10–30[5]	10–20	30–60	
	I	30–60	10–50	50–100[4]	50–120[6]
Soybeans	D	0 or 10–20[3]	10–30	25–50	
	I	0 or 10–20[3]	5–50	50	
Lupins	D	–[3]	5–20	25–50	
Sunflower	D	10–50[5]	5–15	20–40	
	I	15–120	10–30	25–50	20–40
Safflower	D	10–30[5]	5–15	20–40	
	I	30–100	5–20	25–50	
Cotton	I	60–150	10–40	25–50	20–40
Canola (rapeseed)	D	25–75[5]	10–20	30–50	
	I	50–100	15–30	30–50	
Linseed (flax)	D	10–40[5]	10–15	30–50	
	I	20–80	10–25	50	

[1] Rates are given for crops known to be deficient or expected to need fertiliser.
[2] Be guided by the soil P test value in choosing an appropriate rate. The highest rate given is for high rainfall conditions and very low soil P test values.
[3] Nitrogen fertiliser should not be needed if legumes such as clovers, lucerne, soybeans or lupins are effectively nodulated. A small application of 5–10 kg N/ha may assist a struggling young stand of lucerne or lupins but it can also stimulate weed growth. A strip application of 30 kg N/ha will prove nitrogen shortage caused by ineffective nodulation or a deficiency of molybdenum.
[4] A higher rate may be needed where cutting for hay or silage occurs. Add 60 kg K/ha for every crop of hay or silage removed from the site.
[5] Optimum nitrogen rates are influenced by the previous cropping history. For example, less may be needed for a crop following a clover-based pasture while the highest rate could be needed in lighter soils following cereal cropping.
[6] For grazing oats, apply up to 50 kg N/ha after each grazing to a total of 120 kg N/ha.

The greatest nitrogen need for high demand crops, such as maize, comes in the latter half of the growing period, so part of the nitrogen is best applied at this time as a side dressing to maintain supplies throughout the growth period and reduce leaching losses. Foliar sprays of 2% urea at high volume are sometimes used to supplement or replace side dressings.

Table 5   Fertilisers commonly used on pastures and field crops

Fertiliser	Nutrient content	Kg of fertiliser needed to supply 1 kg of N, P, K or S	Advantages	Problems
Urea	46% N	2.17	High analysis – low freight, low cost.	High gaseous losses if left on surface. Water or cultivate in.
Ammonium nitrate	34% N	2.94	Low cost. Quick response. Half N as nitrate which moves quickly to the roots.	Nitrate can leach easily.
Ammonium sulphate	21% N 24% S	4.76 for N 4.16 for S	Not recommended except for alkaline soils.	Expensive. Highly acidifying.
Diammonium phosphate (DAP)	18% N 20% P	5.55 for N 5.00 for P	Supplies both N and P.	Strongly acidifying. Concentrated source of N and P but lacks calcium.
Single superphosphate	8.8% P 11% S	11.36 for P 9.1 for S	Supplies P, S and calcium.	Relatively low P analysis.
Double superphosphate	17% P 4.5% S	5.88 for P 22.2 for S	Concentrated source of P giving savings in freight and handling.	
Triple superphosphate	19% P 2% S	5.3 for P 50 for S	Concentrated source of P.	May lack sufficient S for pastures and some crops.
Sulphur-fortified superphosphate SF 45	5.5% P 43% S	18.2 for P 2.3 for S	For sulphur-deficient pastures which also need some P.	Contains less P than other superphosphates.
Gypsum	12–18% S	8.3–5.6 for S	All S quickly available.	Contains no P.
Potassium chloride (muriate of potash)	50% K	2.0	Cheapest form of K. Highly soluble.	High chloride content. Do not use on saline soils.
Potassium sulphate	41% K 16% S	2.44 for K	Free of chloride.	Higher cost compared to muriate of potash.
Potassium nitrate	38% K 13% N	2.63 for K 7.69 for N	Supplies both N and K in soluble form.	Expensive.
N–P–K mixtures	Various proportions of N, P, K and often S		A convenient way to supply N, P and K in one application.	More expensive than single element fertilisers.

| Blood and bone[1] | 5% N
4% P | 20 kg for N
25 kg for P | Supplies both N and P in slowly available forms. | Large quantities needed. Expensive. |

[1] Typical analysis: the analysis of blood and bone varies from brand to brand.

1

2

1 Perennial ryegrass – response of newly sown grass to fertiliser nitrogen seen as stronger growth and darker green colour (right).

2 Perennial ryegrass – small areas of dark green grass showing better growth (urine patches) within a paddock of pale stunted grass often indicates nitrogen deficiency. The effect was reduced by an application of fertiliser nitrogen to pasture on the left of the fence.

3 Sudax – stunted pale green seedling. Note the purplish colours developing in the older leaves (leaf N = 2.0%).

4 Subterranean clover – pale green to yellow and reddish colours in older leaves of nitrogen-deficient plants resulting from poor root nodulation (leaf N = 1.2%). Clover plants whose roots are nodulated with an effective strain of the Rhizobium bacterium are able to fix adequate nitrogen from the air.

5 Subterranean clover – one dark green healthy nodulated clover plant surrounded by nitrogen-deficient plants which were not nodulated. Nodulation of clover sown into new pasture ground can be assured by innoculating the seed with an effective strain of Rhizobium and, where the soil is acid, by using lime-coated seed or drilling the seed with agricultural lime.

6 Subterranean clover var. Dwalganup – chlorosis, reddening and death of the older leaves. Petioles have turned red. (J. Hirth)

7 Barrel medic var. Jemalong – old leaves have lost nearly all their green colour and developed purplish colours which are strongest near the leaf margins. (J. Hirth)

8 Oats – stunted growth and pale colour of the oats on the left which received no nitrogen fertiliser.

9 Maize – pale green narrower leaves of the nitrogen-deficient plant on the right (leaf N = 1.4%), compared to the large dark green leaves of the healthy plant on the left (leaf N = 3.0%). In this quick tissue test, the white tile shows a dark blue colour produced by nitrate-rich stem tissue from the healthy plant on the left. Compare this to the brown colour from the nitrate-starved stem tissue taken from the deficient plant on the right.

4

5

3

6

10 Oats – the young leaves are a uniform pale green while the older leaves become yellow then red, and finally die prematurely from the tips (leaf N = 1.0%).

11 Wheat – younger leaves of the deficient plant (right) are pale green, while older leaves (centre) become yellower from the tip (leaf N = 1.5%). Healthy leaves are on the left (leaf N = 3.2%).

12

13

14

15

12 Field pea – progression of chlorosis from the pale green upper leaves to reddish pigments in mid stem leaves to yellow lower leaves. In an acute deficiency the lowest leaves may die prematurely. Nitrogen is quite mobile within the plant and moves from older tissues to new growth in times of shortage.

13 Rapeseed – leaves are pale but also develop a reddish purple colour progressing inward from the margins (leaf N = 1.5%). Turnip and other members of the crucifer family develop similar purpling when nitrogen deficient.

14 Field pea – pale green, stunted growth of unnodulated nitrogen-deficient plant on the left (leaf N = 1.4%) compared with the healthy dark green plant at right (leaf N = 3.2%). Note how the margins of older leaves and stipules are tinged with purple.

15 Sunflower – smaller pale green leaves of the stunted nitrogen-deficient plants (left) compared to normal growth and leaf colour and size of plants receiving nitrogen fertiliser (right).

PHOSPHORUS

Many soils in their original unfertilised state have low levels of available phosphorus, and crops and pastures sown into such soils are very often deficient unless adequately fertilised with phosphorus in the first season. Regular applications annually, or with each crop, are usually needed for several years or perhaps indefinitely to maintain adequate supplies of this vital element in the soil. For, although phosphorus is not leached like nitrogen (except from very sandy soils), it becomes more strongly held in the soil and less available to plants with the passing of time since the last application of fertiliser. This process is often called 'fixation' and is most rapid in strongly acid loams and clay loams. In cereal crops, significant amounts of phosphorus are removed in the grain. For example, it has been estimated that half of the normal application of phosphorus to a wheat crop is removed in the grain harvest.

Function   Phosphorus is important for cell division and growth. It is needed for photosynthesis, sugar and starch formation, in energy transfer, and for the movement of carbohydrates within the plant.

Symptoms   A shortage of phosphorus will stunt the growth of plant tops and roots. Growth of seedlings may stop soon after emergence, and cotyledons yellow, shrivel and fall off. In cereals, such as wheat or oats, tillering is curtailed and there is early senescence. Leaves usually have a dull, lustreless appearance. Their colour may become pale or even yellow, especially the older leaves (soybeans). Intense purple or reddish purple colours can develop (sorghum and maize), while in subterranean clover young leaves become a darker green than normal. Phosphorus-deficient linseed produces dull blue-green leaves which are stiff, erect and clasp the stem. Lower leaves wither and die prematurely.

Corrective measures   Plants with obvious deficiency symptoms are most often those sown into new ground. Phosphorus does not leach from most soils and phosphorus residues left from previous fertilising can build up after years of regular fertiliser application. However, loss of availability will occur, especially in acid soils, so most soils need some regular phosphorus fertiliser to maintain high yields.

Phosphorus applications of 10–50 kg P/ha are commonly used depending on the crop, soil fertility, past fertiliser history, soil P test value and various other factors (see Table 4). Superphosphate, ammonium phosphates (MAP or DAP) and compound N–P–K fertilisers are the usual means of supplying phosphorus to pastures and field crops (see Table 5). When fertilising crops, all the phosphorus must be applied at or before planting. Surface applied phosphorus does not move easily through most soils, and so side dressings are ineffective in getting phosphorus to the roots of a growing crop. Pastures, however, are able to benefit from periodic maintenance surface applications of phosphorus because of their greater surface root development.

1 Subterranean clover – when establishing new pasture in previously unfertilised soil, success or failure often depends on quickly overcoming phosphate deficiency with an initial generous application of superphosphate. The very poor growth of both clover and grass in the central plot is due to a lack of phosphorus.

2 White clover – phosphorus has a major role in promoting plant growth. The plot (front right) received no phosphorus while three other plots (front left, middle left and middle right) received increasing rates of phosphorus (12, 24 and 48 kg P/ha respectively).

1

2

3 Subterranean clover var. Dwalganup –
phosphorus-deficient clovers are stunted and
produce small dark green leaves (left). Leaf markings
are more pronounced and the petioles become
reddish purple. (J. Hirth)

4 Barrel medic var. Cyprus – the youngest leaves
are small and dark green, while old leaves turn pale
and yellow, and senesce prematurely. (J. Hirth)

5 Barrel medic var. Jemalong – severe deficiency
with most leaves showing premature yellowing from
the margins inward. (J. Hirth)

6 Berseem clover – phosphorus-deficient plant on
the left (leaf P = 0.14%) is stunted, and has narrow
leaves which show a strong purplish colour when
mature. Healthy plant at right (leaf P = 0.19%).

8

7

9

10

11

12

13

7 Oats – growth is stunted and older leaves develop reddish orange colours commencing from tips. This reddening progresses back along the leaves to the base. Finally these leaves die and this premature senescence moves up the plant affecting younger leaves. Leaf sheaths and stems become reddish purple (leaf P = 0.09%).

8 Maize – strong purple coloration of the leaves characterises phosphorus deficiency in many varieties. Symptoms develop first in the older leaves but quickly progress to young leaves if the deficiency is severe and persists.

9 Maize – Purpling in this variety is mostly along the leaf margins. In other varieties, phosphorus deficiency produces yellow, brown and purplish streaks mainly along the margins and at the tips of leaves prior to tissue death (leaf P = 0.15%).

10 Maize var. XL 45 – severe stunting of maize plants on the left without phosphorus compared to vigorous growth in plants (right) receiving superphosphate banded at sowing. Note this variety does not show leaf purpling in deficient plants but older leaves are seen yellowing and dying prematurely.

11 Sunflower – Leaves are small and a dull dark green. Older leaves then develop dark grey to brown or black necrotic lesions which quickly spread over the distal (tip) half of the leaf.

12 Clover pasture – strips of better growth resulting from the residual phosphorus from fertiliser drilled in rows for a previous crop. Unlike nitrogen fertilisers which quickly leach from the soil, phosphate is retained and can benefit crops for several years.

13 Subterranean clover – Phosphorus toxicity causes burning of margins of outer leaves. This condition is not common in the field, but can occur on very poor sands that are low in nitrogen and organic matter when high rates of phosphorus (greater than 40 kg P/ha) are applied in a single dressing.

 POTASSIUM

Most soils contain large amounts of potassium, but only a small proportion of it (mostly the exchangeable fraction) is readily available. The remaining reserves are released more slowly and, if peak crop or pasture demand exceeds the rate of release, plant production will suffer and potassium deficiency symptoms appear.

Potassium deficiencies are most likely to occur on lighter soils in high rainfall areas. Heavier soils usually have greater supplies and reserves of potassium, but even the richest soils can become deficient under intensive cropping, hay cutting or silage making, where large amounts of potassium are removed in produce. Although grazing animals return potassium in their excreta, grazed areas can still suffer potassium depletion through the redistribution of this nutrient to other parts of the farm, such as sheep camps, night paddocks or holding areas near the dairy.

Potassium can also become limiting when other fertilisers, particularly nitrogen, are used at high rates to increase production. Heavy liming can also induce potassium deficiency by changing the balance of exchangeable potassium to calcium in the soil.

Function   Potassium is important for forming proteins, carbohydrates and fats and for the functioning of chlorophyll and several enzymes. It is necessary for cell division, maintaining the balance of salts and water in plant cells and for opening and closing the stomates – the tiny breathing apertures on the undersurface of leaves. Potassium is highly mobile and can be moved freely within the plant to new tissues where it is most needed. It is found in highest concentrations in leaves, growing points, flowers and fruit.

Symptoms   Scorching of the edge and sometimes between the veins of older leaves characterises potassium deficiency in most crops. In clover and lucerne, small necrotic spots often develop between the veins forming a distinctive speckled pattern, but sometimes the necrosis forms a continuous marginal burn and leaf cupping can occur. Leaf burn can be light brown to almost black in colour and may be preceded by yellowing of leaf margins or interveinal areas of older leaves. Scorching of the leaf margins can cause expanding leaves to curl downward or to cup upward.

In some crops, including maize, barley, sunflower, soybeans and cotton, potassium deficiency leads to a shortening of the internodes which gives a compressed growth habit. Maturity is often delayed, and yield and quality may be reduced even before leaf symptoms are seen. In a mixed pasture, the first indication of potassium shortage may be a decline in the legume content, leading first to grass dominance and then to yellow nitrogen-deficient grass. This happens because grasses are generally better able to compete for the limited potassium supply.

Corrective measures   Potassium chloride (50% K) (muriate of potash) is the cheapest fertiliser source and is generally the one used for pastures and field crops. Potassium sulphate (42% K) contains less potassium and is more costly, as is potassium nitrate (38% K, 13% N) which contains nitrogen as well as potassium (see Table 5). Use of these latter two forms is normally restricted to horticultural crops, except for tobacco. The quality of this high value field crop can be harmed by the chloride in muriate of potash.

Rates of potassium application vary enormously: from zero for a non-irrigated low demand crop, such as wheat sown on a heavy inland soil, to over 100 kg K/ha for a high demand crop like maize on a leached coastal sandy soil with a history of long-term intensive cropping, strip grazing, or fodder conservation. Applications of 30–60 kg K/ha represent average trial rates for a known or suspected deficient pasture or crop site, but where potassium is constantly removed in farm produce, such as lucerne or pastures cut for hay or silage, an additional 50 kg K/ha should be applied for each cut removed.

Except in very sandy soils, leaching of fertiliser potassium is not usually a serious concern. Nevertheless, timing and placement of fertiliser is important. For short-term crops, the whole potassium fertiliser requirement is usually incorporated into the soil at or before sowing, because movement of potassium from surface applications or side dressings may be too slow to benefit a rapidly growing crop. For the same reason, correction of a deficiency in an existing crop is difficult, so the objective should be prevention in future crops. Foliar sprays can be phytotoxic and generally supply too little potassium to save a deficient crop.

In longer term crops like lucerne and in pastures, potassium fertiliser is topdressed and, sometimes split into three or four dressings per season. This avoids waste from luxury uptake into the first cuts of lucerne or pasture hay, and it reduces the risk of mineral imbalance (low magnesium) in the pasture, which can cause grass tetany in cattle.

1 White clover – the most characteristic symptoms are numerous small white to yellow necrotic spots developing just in from the margins of older leaves. Complete necrosis later spreads inwards from the margins until the whole leaf is dead.

2 White clover – close up of a single leaflet showing silvery white necrotic spots between the veins.

3 Lucerne – (left leaf) chlorosis of the outer part of older leaves with some necrotic white to yellow spots developing between the veins within the chlorotic zone. The other leaves (centre and right) have developed more necrotic spots and less chlorosis, (leaf K = 0.6%).

4 Lucerne – because potassium moves from old to new leaves when deficient, symptoms are first seen and develop most strongly in the oldest leaves. Note the youngest fully open leaf is still green, the next is pale green with a few necrotic spots near the margins of individual leaflets, while chlorosis and necrosis increase down the shoot (leaf K = 0.8%).

5 Lucerne – the distinctive tiny spots are not always prominent in lucerne, medics or clovers. Yellowing and death of the outer portions of leaflets of lower leaves are the main symptoms in this specimen (leaf K = 1.4%).

6 White clover – chlorosis followed by dark brown necrosis around the margins and outer parts of the leaflets are the main symptoms here (leaf K = 0.5%). The tiny spots are absent. Often in the field, distinct leaf symptoms may not be noticed at all and the only clue to a developing potassium deficiency is a decline in the clover population. Because clovers are generally less able to compete for a declining soil potassium supply, the balance of clover in the pasture mixture diminishes and the sward becomes a nitrogen-deficient mixture of grasses and weeds.

1

2

3

7 Berseem clover – various stages and forms of leaf chlorosis, small necrotic spots, and marginal and interveinal leaf burn.

8 Barrel medic var. Cyprus – chlorosis followed by necrosis of the outer sections of leaflets of older leaves. (J. Hirth)

9 *Dolichos lablab* – yellowing, progressing inwards between the veins from the leaf margins (leaf K = 0.46%).

10 Soybean – a creamy yellow chlorosis spreading from the tip and margin of the leaflets, followed by a light brown necrosis (leaf K = 0.4%). In some varieties necrosis near the leaf margin and between the veins may be the main symptom with little or no chlorosis. Affected plants are short, with thin stems.

11 Sunflower – chlorosis commencing near the leaf tip (far left), spreading around the margin (centre bottom), and finally intruding to between the veins and moving towards the leaf base (top centre). Light brown necrotic areas then develop from the leaf tip, margins and sometimes the interveinal areas (top centre). The leaf on the right is healthy. Potassium deficiency in sunflowers and several other species, causes a dramatic reduction in stem length giving the plant a squat appearance.

12 Cowpea – chlorosis extending from the tips and margins of leaflets with light brown irregularly shaped necrotic spots in the interveinal areas (leaf K = 0.48%).

13 Maize – a light brown necrosis commencing at the leaf tip and then advancing along the margins towards the leaf base (leaf K = 0.3%). Some chlorosis usually precedes the necrosis but is quickly overtaken by it. Note: yellowing and necrosis extends back along the outside of the leaf (in contrast to the symptoms of nitrogen deficiency, where the yellowing and necrosis follow along the midrib).

14 Maize – plants are shortened and paler than normal; older leaves develop yellow and then necrotic bands commencing from the tips and spreading along the margins. (A. Dale)

15 Barley – chlorosis and light brown necrosis from the tip and margins of older leaves.

9

10

12

11

13

15

14

17

16

16 Peanut – light creamy yellow chlorosis commencing near the tip of leaflets (centre leaf) and progressing inwardly between the veins with marginal necrosis commencing (right leaf). The leaf on the left is healthy. Affected leaves contained 1.1% K.

17 Linseed (flax) – brown necrosis of the tips of lower leaves. (P. Hocking)

SULPHUR

Sulphur is required by plants in roughly the same amounts as phosphorus, yet deficiencies of sulphur in the soil have often gone unrecognised. This is because of the supply of sulphur in fertilisers such as superphosphates, ammonium sulphate or potassium sulphate used to correct other deficiencies. Sulphur is more often deficient in pastures than in crops, although crop responses do occur. Some cruciferous crop plants, including canola (rapeseed) and turnip, have a high requirement. Cultivation releases available sulphur from soil organic matter, making sulphur deficiency less common in crops and first-year pasture sown into cultivated land.

Sulphur deficiency is more likely to be found in some areas and soils than others. In eastern Australia, it is most common in areas of higher altitude some distance from ocean and industrial accretions in rainfall; in western parts of the continent, sulphur deficiency is often found in areas of relatively intense winter rainfall and porous sandy or granitic soils, some of which are near the coast.

Function   Sulphur is a constituent of several amino acids that are essential for building both plant and animal proteins. It is also needed to activate some enzymes and for the synthesis of chlorophyll. A shortage reduces growth and production and lowers the protein content of pasture. It can also cause a loss of baking quality in wheat flour and a lower oil content in some oilseed crops.

Symptoms   These resemble the effects of nitrogen deficiency, with the foliage developing a uniform pale green to lemon yellow colour over the entire leaf blade (the chlorosis has no interveinal or other pattern). However, the symptoms differ from nitrogen deficiency in that the young leaves become yellow (though older leaves may be yellow also). With nitrogen deficiency yellowing always begins with and is most severe in the old leaves.

When sulphur deficiency is severe and persists in clovers, the leaflets of old leaves can become very pale and die from the margins inwards. The nodules produced by sulphur-deficient legumes are smaller, fewer in number, and white rather than a healthy pink colour.

Corrective measures   Many common pasture and crop fertilisers contain significant amounts of sulphur which will correct or prevent sulphur deficiency (see Table 5). Phosphorus is very often deficient in low-sulphur soils. If single superphosphate or even sulphur-fortified superphosphate is used instead of a low sulphur fertiliser, such as triple superphosphate, both deficiencies will be corrected. Likewise, where both sulphur and nitrogen are needed, ammonium sulphate can satisfy the dual need. Other common sulphur fertilisers include gypsum and finely powdered elemental sulphur. The latter has an advantage where intense leaching of sulphate is a problem.

Rates of application can vary from less than 10 kg S/ha to 60 kg S/ha but a common rate for pastures is around 12–15 kg S/ha. For a deficient, high sulphur-demanding crop such as canola (rapeseed) apply 20–30 kg S/ha.

1 White clover – without sulphur (right plot) growth is stunted and plants are pale green. The left plot and those in the middle distance received sulphur.

2 Subterranean clover/phalaris – both the clover and the grass are paler in colour and stunted where sulphur has not been applied (right). *Phalaris aquatica* responds well to sulphur in the field, but its symptoms are a less reliable guide than clover, because they may be confused with symptoms of nitrogen deficiency. Nodulated clover is not dependent on soil nitrogen.

3 Subterranean clover – generally pale with smaller leaves (left). Some of the oldest leaves are turning yellow and senescing.

4 Subterranean clover – healthy leaf (left) with uniformly pale green leaf (centre) and yellow leaf (right). Veins may remain green for some time. Finally, the oldest leaves of sulphur-deficient clover may scorch from the margin and die.

5 *Phalaris tuberosa* – uniformly pale lemon yellow sulphur-deficient leaf (right) compared to a healthy dark green leaf (left).

6 Subterranean clover var. Dwalganup – chlorosis of older leaves. (J. Hirth)

3

4

5

6

7

7 Barrel medic var. Jemalong – in this specimen, the youngest leaves are most chlorotic. Clovers and medics can vary as to whether yellowing is greatest in old or young leaves, depending on the species, the intensity and duration of the deficiency, and whether the legume is wholly reliant on root nodules for nitrogen or receives some nitrogen from the soil or fertiliser. (J. Hirth)

8 Lucerne – smaller, paler green leaves and poorer growth of the two sulphur-deficient shoots (left and centre). Healthy plant on the right.

9 Subterranean clover – The residual effect from fertiliser sulphur is often low due to leaching and other losses. All these plots received 84 kg S/ha three years earlier. An application of 24 kg S/ha was given in the current season to the plots on the right and in the foreground. Yellowing of the clover on the left indicates that little of the original sulphur was still available to the pasture. Leaf S = 0.11% (yellow clover) and 0.20% (healthy clover).

8

9

10 Canola (rapeseed) – mottled yellowing of leaves, especially younger leaves. Faint purple tints develop near the edges of leaves which curl inwards. (H. Burns)

11 Canola – the flowers of sulphur-deficient canola (right) are a pale creamy colour compared to the normal bright yellow flower of a healthy plant (left). Seed set is poor, pods abort or are short and bulbous. Deficient plants continue to flower long after the healthy crop has ripened and is ready for harvest. (H. Burns)

10

11

CALCIUM

Soils that are low in calcium usually have other associated problems including extreme acidity and harmful levels of soluble aluminium and/or manganese. Sandy soils which have been heavily leached by high rainfall and acid peat or muck soils are among the lowest in calcium.

Calcium deficiency, as distinct from soil acidity problems, is less common in pasture and field crop plants than it is in horticultural plants where a shortage of calcium is often expressed in classic disorders like bitter pit in apples or blossom-end rot of tomatoes. Even in leached acid soils, the calcium needs of field crop and pasture plants is often met by the calcium (22% Ca) in superphosphate.

Flax is one of the more susceptible field crops which sometimes develop symptoms of calcium deficiency, described as 'withertop'. This usually occurs in the spring, especially in waterlogged acid soils. In plants which form their seed underground (peanuts and subterranean clover), low soil calcium can result in a failure of the seeds to form.

Function   Calcium is a main constituent of the cell walls and membranes and, when it is in short supply, cell membranes become leaky, and cell division and the development of the growing point and root tips are affected. Calcium activates several plant enzymes, stimulates nodulation in legumes, and has been shown to have a role in nitrogen fixation in subterranean clover.

Symptoms   Because calcium has poor mobility within the plant, new tissues are the first to be affected by a deficiency. Death of the growing point or areas of tissue breakdown in the root tips, young leaves, or developing fruit are common expressions of calcium deficiency. Young emerging leaves of grasses and cereals do not unfold but remain rolled and stick together at their tips. The petioles of young clover and lucerne leaves can collapse suddenly, as can the flowering stems of flax and canola, causing death of flowers and the ends of the stems.

Corrective measures   Foliar sprays of calcium nitrate or calcium chloride (800 g/100 L) are not usually appropriate treatments for field crops or pasture plants. Where calcium deficiency symptoms are evident, the soil is probably quite acid and will usually require lime to overcome other problems such as aluminium or manganese toxicities. Agricultural lime, which contains 35–38% calcium (Ca), broadcast at rates of 500 kg/ha to several tonnes per hectare to correct soil acidity, supplies significant amounts of calcium to the soil.

Gypsum (calcium sulphate), containing 17–23% Ca, is another good source of calcium. It does not correct soil acidity, but it is used to improve the structure and help remove sodium from saline and sodic soils.

2

1

3

5

4

7

8

9

1 Subterranean clover – in clovers and medics calcium deficiency causes sudden collapse of petioles of young leaves. The leaf is normal at the time of the petiole failure but quickly withers and dies.

2 Subterranean clover – general petiole collapse followed by leaf death observed in plants grown without calcium fertiliser in a low calcium soil. Superphosphate and lime applications often supply sufficient calcium to prevent symptoms in field-grown clovers. (J. Hirth)

3 Subterranean clover var. Mt Barker – a wine red colour often develops on leaves, particularly the undersides of leaves, and on petioles. (J. Hirth)

4 Subterranean clover var. Dwalganup – reddish purple pigments on the leaves. (J. Hirth)

5 Lucerne – collapse of the stem and the petioles of young terminal leaves followed by leaf death.

6 Barrel medic var. Cyprus – reddening around the outer leaflet and chlorosis of the central areas. (J. Hirth)

7 Canola (rapeseed) – mild 'withertop' where the flowering stems collapse causing a failure of pods to form and seeds to set.

8 Canola – a more severe form of 'withertop' involving complete death and browning of the flowerhead. (R. Colton)

9 Linseed (flax) – collapse of flower stalks just below the buds is described as 'withertop'. (P. Hocking)

MAGNESIUM

Although magnesium deficiency is mostly thought of as a problem of horticultural crops, it does affect a number of field crops, including wheat, maize, tobacco, cotton and oats, when grown on acid sandy soils. Pasture plants rarely show deficiency symptoms or have their growth affected in the field by low soil magnesium, but livestock, particularly dairy cattle, can be seriously affected by grass tetany when the magnesium level in the pasture they graze is low.

Clay soils, especially those which are neutral or alkaline in pH, are usually well endowed with magnesium. Soils which are most likely to lack sufficient magnesium are acid sandy soils that have been strongly leached, acid peats, and soils which have become very acid through extended heavy use of ammonium fertilisers for intensive cropping. Soils formed from sandstone are commonly deficient, but the basalt-derived red loam soil (kraznozem) of high rainfall areas is also often lacking in magnesium. Liming or fertiliser programs which supply high inputs of calcium or potassium can change the soil's nutrient balance over time and induce magnesium deficiency, unless some magnesium is provided through judicious use of dolomite or magnesite.

Function Magnesium is present in chlorophyll, the vital green pigment which enables plants to form sugars and starches from atmospheric carbon dioxide. Several plant enzymes, particularly those associated with photosynthesis, need magnesium to function.

Symptoms Magnesium is quite mobile within the plant, so deficiency symptoms usually appear first, and become most severe, in the oldest leaves. The first sign is usually a slight paleness in these leaves, but a patterned chlorosis soon develops with tissue close to the midrib and major veins remaining green. The interveinal chlorosis can be pale green to bright yellow or occasionally orange or bronzed. Necrotic areas develop within the chlorosis in many plants including maize, cotton and soybeans. Because of the parallel venation in leaves of maize, cereals and grasses, chlorosis gives the leaves a striated appearance. In wheat and oats, chlorosis can form longitudinal 'beading' between the veins. In the final stages, the margins and tips of chlorotic leaves may die.

Corrective measures The main materials used to correct a deficiency of magnesium are the liming materials, dolomite (calcium magnesium carbonate 8–13% Mg), magnesite (magnesium carbonate 20–28% Mg), magnesium oxide (40–55% Mg), and the soluble salt magnesium sulphate or Epsom salts (9.6% Mg). If you know the soil is deficient prior to planting, a preventative application of 100–200 kg magnesium oxide, up to 500 kg magnesite or up to 1–2 tonnes of dolomite per hectare should eliminate the problem for several years. Response in a growing crop or pasture may take many months where treatments are topdressed. Foliar sprays of 2% (2 kg/100 L) magnesium sulphate (Epsom salts) at high volume (500–1000 L/ha) are usually needed for a quick correction of an existing deficient crop.

1

2

3

1 Wheat – chlorosis of the leaves with the oldest leaves becoming necrotic and dying from the tip and margin. The first fully expanded leaf (upper left) has developed a 'beaded' interveinal chlorosis (leaf Mg = 0.09%).

2 Wheat – 'beaded' striping is characteristic of magnesium deficiency in cereals, particularly wheat and barley (leaf Mg = 0.08%). 'Beading' may not always be found in the field. Yellowing and dying of the older leaves from the tip is often the only symptom, making it difficult to differentiate this disorder from potassium or nitrogen deficiencies.

3 Wheat – patchy yellow stunted growth in a crop caused by low magnesium uptake from a highly acid soil.

4

5

7

6

4 Maize – interveinal (striped) chlorosis of the older leaves, developing from the tip and mid sections of the leaves and advancing towards the leaf base (leaf Mg = 0.16%).

5 Maize – in the advanced stages of deficiency, necrotic lesions develop in the chlorotic areas, and may spread along the lower leaves until they collapse and die.

6 Subterranean clover var. Dwalganup – red coloration begins around the edges of older leaves and moves inward. Leaf margins then become necrotic. (J. Hirth)

7 Subterranean clover var. Dwalganup – with continued severe deficiency leaf necrosis moves steadily in from the leaf margin. (J. Hirth)

8

9

10

8 Barrel medic var. Cyprus – yellow chlorosis starting at the apical margins of older leaves. Note the young leaves are still green. (J. Hirth)

9 Barrel medic var. Cyprus – as deficiency progresses, chlorosis extends in from the margin towards the centre and small dark brown necrotic spots develop in the bleached areas. (J. Hirth)

10 Sunflower – interveinal mottled yellowing of older leaves. The tissue near the midrib, major veins and leaf base remains green longest (typical of magnesium deficiency). In the oldest leaves, light brown necrotic patches later develop in the chlorotic areas (left leaf).

11

12

11 Cotton – mottled chlorosis developing between the veins of older leaves, with brown necrotic spots forming in the chlorotic areas.

12 Oats – yellow strips of magnesium-deficient oats. These rows coincide with bands of heavy potassium fertiliser applied to a prior crop of potatoes. Magnesium deficiency is often induced by nutrient imbalance, especially following heavy applications of potassium, calcium or ammonium fertilisers.

*B*oron deficiency is a common disorder of many field and forage crops, and of lucerne and pasture legumes. The deficiency is most often found in tableland districts, especially in sunflower and canola crops, and in lucerne and clovers. Except for maize, which sometimes develops deficiency, most cereals and grasses tolerate moderately low levels of boron in the soil. Table 6 groups some common crop and pasture plants according to their sensitivity to boron deficiency.

Table 6   Sensitivity of some crop and pasture plants to boron deficiency

Susceptible	Moderately susceptible	Tolerant
Lucerne	Clover, white	Grasses
Sunflower	red	Barley
Rapeseed	subterranean	Oats
Turnip	Persian	Wheat
Sugarbeet	Townsville stylo (*Stylosanthes humulis*)	Rye
Cotton	Tobacco	Rice
	Maize	Sorghum
	Sweet corn	Soybean
		Safflower

Soils vary in their capacity to supply boron. Light-textured soils (such as granites) which are low in organic matter, especially those in high rainfall areas, are more prone to deficiency than heavier textured soils in drier areas. Soils derived from marine sediments often have the highest reserves of boron. Unlike other trace elements, boron is easily leached from the soil. Its availability to plants is also reduced during dry periods. Deficiency often develops when a long wet period is followed by a dry spell. Because plants require a small but continuous supply of boron from the soil, symptoms may suddenly appear even though the problem has not been seen for a number of years. Heavy liming also reduces boron uptake.

Function   Boron is important for cell division and organisation in the growth regions of the plant, that is, near the tips of shoots and roots. It has a role in the metabolism of auxin, an important growth hormone, and is needed for moving sugars within the plant. Boron is required for pollination and the development of viable seeds.

Symptoms   Tissues are brittle and crack or split easily. The surface of stems, petioles, or the midribs of leaves may become corky or crack, storage roots split, or stems develop hollow sections. Boron does not easily move around the plant, so deficiency symptoms first appear and are usually most acute in young tissues. The ends of shoots may become shortened (for example, umbrella-shaped growth in lucerne), or strongly distorted (sunflower), or the growing point may die leading to multiple crowns (beet and sunflowers). Young leaves can develop yellow to orange tints (canola, lucerne), or red and purple coloration (clover), or the leaves may become distorted, thickened or leathery. Common but sometimes less noticed effects of boron deficiency include reduced flower numbers, low pollen production, poor seed set (clovers, lucerne, rapeseed), and sterile florets. Root injury or stunting can occur, but this often remains unseen in early deficiency.

Corrective measures   Once deficiency symptoms are evident, the damage can be serious and permanent in a crop such as sunflowers. Therefore, a preventative soil application of 1–3 kg B/ha (10–30 kg/ha borax) at or before sowing is preferable. However, a foliar spray of 2–10 kg/ha of polyborate powder used early in a crops growth, before severe symptoms develop, can prevent serious damage. Rates vary with crop sensitivity, demand and soil type. Table 7 gives recommended rates for a range of crops.

Table 7   Boron application rates for common forage and field crops

Crop	Boron (B) (kg/ha)	Soil application borax[1] (kg/ha)	Foliar application polyborate powder (e.g. Solubor) (kg/ha)	Maximum concentration (% w/v)
Lucerne	1–3	10–30	5–10	0.5
Clover	1–1.5	10–15	5–10	0.5
Sunflower	1–3[2]	10–25	5–15	0.25
Rapeseed	1–2	10–20	5–12	1.6[3]
Linseed	0.5–1	5–10	2–5	0.25
Cotton	0.5–1	5–15	2–10	0.5
Turnip	1–3	10–25	5–10	0.5
Sugar beet	1–3	10–25	5–15	2.5[3]
Maize (and sweet corn)	0.5–1	5–10	2.5	0.5

[1] If using polyborate powder (Solubor) for soil application, halve rates given for borax.
[2] Use the lower rate if applying borated fertiliser in a band.
[3] These maximum concentrations apply to low volume sprays only. Concentrations for high volume sprays should not exceed 0.25–0.5% w/v.

1 Lucerne – yellow-orange leaves with shortened growth near the top of the deficient shoots (centre and right) compared to healthy normal growth (left). Note that the lower leaves on affected shoots are green.

2 Lucerne – paleness and reddish coloration of upper leaves. The top of the shoot fails to lengthen normally, giving it an umbrella shape (leaf B = 10 ppm).

3 Lucerne – the upper leaves often become strongly coloured with bright red, yellow and orange hues (leaf B = 13 ppm).

4 Lucerne – when boron deficiency is mild or comes late in the season, the only visible symptom in lucerne (and other crops) may be a failure of the flowers to set seed (leaf B = 16 ppm).

5 Subterranean clover – reddening of the outer regions of mature leaflets. Slight marginal necrosis is also evident. Necrosis is commencing along the margins of some leaflets. Petioles are often reddish in colour and shortened. Leaflets tend to close by folding inward.

6 Subterranean clover – a range of red and yellow colours develop, especially towards the edges of leaflets. Reddening is usually strongest on the undersides of leaves (top left). (B. Dear)

7 Sugar beet – cracking of the root, which is hollow in its centre, leading to 'heart rot'. The leafy top becomes a mass of many small shoots ('multiple crowns') as the main crown fails.

8 Berseem clover (*Trifolium alexandrinum*) – purpling of the outer portion of leaflets (leaf B = 17 ppm).

9 Sunflower – because boron does not easily move from old to new growth, deficiency may suddenly affect upper leaves and the flower, though early growth appeared normal. Upper leaves of the deficient plant (right) have failed to expand and are grossly distorted, with necrosis developing near the leaf base. The flowerhead is distorted and barren, and leans to one side because part of the flower stalk has died.

10 Sunflower – when the deficiency is severe, terminal growth stops, the upper leaves yellow, twist and die, and the flower fails to emerge from the bud.

4

5

7

6

8

9

10

12

11

11 Sunflower – greyish brown necrotic areas form in interveinal tissue, mostly near the central vein and towards the base of leaves.

12 Sunflower – the effect of severe boron deficiency on the youngest tissues, death of the growing point which is surrounded by small deformed upper leaves. Lower leaves are almost normal in size and shape. Failure of the main growing point has induced secondary shoots, which will also fail.

13 Canola (rapeseed, _Brassica napus_) – yellowing of leaves with some orange colour developing. (B. Dear)

14 Canola – reduced flowering and failure to set seed are serious effects of boron deficiency. They can occur late in the crop's development without prior leaf coloration being evident. (B. Dear)

15 Maize – broken creamy yellow streaks on the leaves.

16 Maize – yellow streaks on the leaves resulting from boron deficiency can be less distinctly patterned than those shown by the previous specimen, when they are easily confused with symptoms caused by some viruses.

MOLYBDENUM

Molybdenum is less available in acid soils and deficiency occurs widely on these soils in the higher rainfall areas of south-eastern Australia. However, the disorder is also found in other areas where soils, once near neutral in their pH, have become increasingly acid under cultivation and pasture development.

Molybdenum deficiency has been a major problem in the establishment of legume-based pastures. Subterranean, white and red clovers and lucerne are most often affected, especially in new sowings, or where molybdenum has not been applied for many years, or where soil pH has fallen. Crops, including maize, sunflower, canola, and occasionally wheat, are also affected.

Function   Molybdenum is needed in chemical changes associated with the nitrogen nutrition of plants. It is required to reduce nitrates as a first step to forming proteins and other essential components, including enzymes and chlorophyll. As a result, deficient plants develop symptoms of nitrogen shortage (paleness and stunting). However, in addition to this, burns develop on the edges of leaves where nitrates, which the deficient plant cannot metabolise, have accumulated. Molybdenum is required by the bacteria in the root nodules of legumes to fix nitrogen from the air.

Symptoms   In lucerne, clover and other pasture legumes, the main symptoms result from their inability to fix atmospheric nitrogen. The stunting and yellowing is identical with nitrogen deficiency and the plants resemble legumes lacking effective nodules and growing in poor soils. Deficient non-leguminous plants, like maize or sunflowers, are also stunted and pale but they also develop a burn around the leaf margins or tips. Applying nitrogen fertiliser to pale, molybdenum-deficient non-legumes will usually accentuate the symptoms, particularly leaf burn.

Corrective measures   Liming the soil to raise the soil pH to about 6.0 (1:5 water) or higher will increase the availability of molybdenum in the soil, but direct addition of molybdenum to the soil or crop is usually the cheapest and most reliable treatment. For clovers, lucerne or other pasture legumes, apply molybdenum trioxide at 75 g/ha (or equivalent amounts of sodium molybdate or ammonium molybdate) mixed with superphosphate. Under non-cultivated pasture conditions, a single application usually remains effective for 3–5 years.

For crops such as sunflowers, field treatment at planting with 250–300 g/ha of sodium molybdate or ammonium molybdate will ensure adequate molybdenum for the crop, but re-treatment may be needed for subsequent crops. Foliar sprays applied early enough in a crop's growth can correct a deficiency within a week and the crop may recover fully, but spraying the crop after the onset of symptoms may not be effective in restoring full yield potential. A spray of 0.05% (50 g/100 L) of sodium molybdate or ammonium molybdate should be used on young crops.

Because molybdenum is required in relatively small quantities, the amount stored in the seed of some large-seeded crops, such as maize, can prevent deficiency symptoms from developing, provided the seed comes from a seed crop which had a plentiful molybdenum supply. Treatment of hybrid maize seed crops with two sprays of sodium molybdate (50 g/100 L/ha) at 30 cm and 80 cm crop heights can prevent molybdenum deficiency in crops sown with seed from seed crops treated in this way, even when it is sown into deficient soils (Weir *et al.* 1976)[*].

[*] Weir, R.G., Nagle, R.K., Noonan, J.B. and Towner, A.G.W. (1976). The effect of foliar and soil applied molybdenum treatments on the molybdenum concentration of maize grain. *Aust. J. Exp. Agric. An. Husb.* **16**: 761–4.

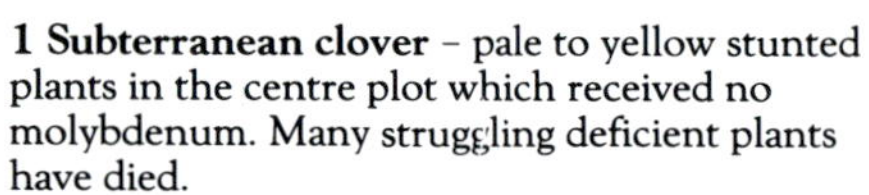

1 Subterranean clover – pale to yellow stunted plants in the centre plot which received no molybdenum. Many struggling deficient plants have died.

2 Lucerne – pale, stunted and less leafy plants on the left without molybdenum compared to healthy molybdenum-treated lucerne on the right. Both plots had received 2 tonne/ha of lime.

3 Sunflower – pale stunted deficient plant (left) compared to a normal plant (right). Burning of the margins has led to cupping of the developing leaves.

4

5

6

7

8

4 Sunflower – molybdenum deficiency in sunflower begins as paleness of the leaves of the affected plant (right) compared to the dark green leaves on the plant at the left. Leaves are pale because deficient plants are unable to use the nitrates they absorb from the soil to form proteins, chlorophyll and other vital substances.

5 Sunflower – a later stage of deficiency where the leaves have become yellow and unused nitrates have caused burns around the edges and between the veins of older leaves. A sap test of tissue near the burnt margins will usually show very high levels (5000–20 000 ppm) of nitrate-nitrogen.

6 Maize – molybdenum deficiency affects maize most severely in the seedling stage. Affected seedlings, like the one on the left, are pale, stunted and often distorted, and many die at this early stage.

7 Maize – deficient seedlings are pale and develop leaf burn, especially near the tips which often fail to unroll.

8 Maize – the boundary between healthy and necrotic tissue often has a watery translucent appearance. Healthy seedling at right.

9

10

12

11

9 Wheat – deficient plants (right and centre) are pale and appear slightly wilted, and are less erect than normal. Healthy green plant at left. Mild deficiency in the field is not easily recognised from symptoms even when yield losses are significant.

10 Wheat – severe deficiency, especially when soil nitrates are high after a long fallow or heavy fertiliser use, results in scorching of the leaf tip and interveinal mid-leaf tissue. This is caused by the accumulation of non-metabolised nitrates in the leaves.

11 Canola (rapeseed) – leaves are very pale with slight mottling of the interveinal areas and cupping of the leaf margins (leaf on the right).

12 Peanut – even pale green to yellow leaves. Some of the older leaves have retained their colour.

*I*ron deficiency is sometimes called 'lime-induced chlorosis' because it often causes chlorosis in crops grown in calcareous soils where the high pH renders iron unavailable. Free lime concretions near the soil surface, a low soil oxygen supply to the roots caused by a high watertable, or low soil temperatures can all add to the problem. Excessive lime applications can also reduce iron availability in the soil and induce a deficiency.

Function   Plants need iron to produce chlorophyll and to activate several enzymes, especially those involved in the oxidation/reduction processes of photosynthesis and respiration.

Symptoms   The youngest leaves develop a light green chlorosis of all the tissues between the veins. This produces a distinctive pattern formed by the midrib and veins which initially remain green. If the condition is severe and persistent, chlorosis becomes more yellow or even white and burnt patches develop within these chlorotic areas. Because iron does not move easily within the plant, older leaves can remain completely green while new leaves emerge which are progressively more and more chlorotic.

Corrective measures   Applications of iron salts to the soil are often ineffective because they change to insoluble forms unavailable to plants but, in sandy soils, dressings of 50–100 kg/ha of iron (ferrous) sulphate have proved effective. However, residual effects cannot be expected beyond the season of application. Iron chelates can be more effective and longer lasting, especially in heavier soils, but they are usually too expensive for field crops or pastures.

Foliar sprays of 1–2% iron sulphate (1–2 kg/100 L/ha) are a cheap and effective alternative to soil treatments, but iron is not very mobile within the plant, so repeat sprays at 2-weekly intervals may be required to provide iron for new growth.

Sometimes a change in management practices, such as attention to drainage, careful use of irrigation to avoid over-watering, the selection of more tolerant cultivars or species, or the use of acidifying fertilisers such as ammonium sulphate, provide the only economic and best long-term solution to this problem deficiency.

1

2

3

1 Sorghum – pale green to yellow interveinal chlorosis of the younger leaves. All leaves in this photo show some chlorosis, but those in the inner whorl (youngest) are the most chlorotic. Sorghum is relatively sensitive to iron deficiency.

2 Maize – chlorosis between the veins of young leaves. Because iron does not move readily from old leaves to young, the older leaves are often completely green while newly formed leaves are progressively more chlorotic.

3 Subterranean clover – chlorosis of young leaves with just the veins remaining green. In more severe cases, the whole leaf can become yellow to almost white and necrotic areas develop at the centre and bases of leaflets.

4

5

7

6

4 Chickpea – interveinal chlorosis, increasingly severe towards the uppermost leaf. The lowest leaves are a normal dark green colour.

5 Field pea – the youngest leaves are completely pale, including all veins.

6 Lupin – increasing chlorosis, from lower normal green leaves, through pale green first fully expanded leaf (upper right) to yellow unexpanded youngest leaves. Deficiency symptoms are most severe in the youngest leaves because iron is not very mobile within the plant.

7 Lupin seedling – chlorotic young leaves except for the midrib which is still green at this stage. The cotyledons have remained green because seed reserves were initially adequate and iron mobility within the plant is poor.

 MANGANESE

Manganese deficiency is not a common or serious disorder of field or pasture crops in New South Wales – toxicity is far more prevalent. However, in southern and western areas of Australia, and in other parts of the world including the United Kingdom, the United States and Europe, serious deficiency has been reported, particularly in crops of field peas, lupins, soybeans and oats, and sometimes lucerne and pasture legumes.

Plants differ in their susceptibility to manganese deficiency. Oats, wheat, field peas, soybeans, sugar beet and canola are most sensitive while maize, rye and pasture grasses are more tolerant of low manganese supply. Barley, lucerne, clover and rice occupy an intermediate position in this scale of sensitivity. There is also a wide diversity in the sensitivity of different cultivars in many of these crops. Manganese is normally less available in calcareous, high pH soils, and deficiency can be induced by over-liming soils which were originally acid.

Function   Manganese is necessary for chlorophyll production, for photosynthesis, respiration, nitrate assimilation and for the activity of several enzymes.

Symptoms   Manganese deficiency often develops as pale patches of poor, stunted plants surrounded by healthy normal crop. Symptoms may be evident early in the crop's development, or appear late in the season when the availability of manganese has declined through drying out of the soil during a period of dry weather.

In wheat, symptoms first appear in the **young** leaves as paleness and wilting. White to grey streaks develop between the veins towards the leaf base. Severe deficiency causes death of the youngest leaves and sometimes of whole shoots. In oats, which is usually more sensitive to low soil manganese than wheat, the **old** leaves show the symptoms first and are the most affected. Grey to brown flecks develop about halfway along these leaves. The flecks join to become large lesions (described as 'grey speck') causing the leaf to collapse at one of these weakened positions. Lucerne develops a faint chlorosis of the interveinal areas which becomes more distinct if the deficiency worsens. A few brown necrotic spots become scattered over the affected leaves.

The seed shows distinctive symptoms in field peas and lupins. The leaves may show only a mild mottling, while the seeds are seriously affected. Shortage of manganese causes a 'split seed' condition in sweet narrow-leafed lupin, in which the seed opens along its edge furthest from the point of attachment to the pod and the cotyledons protrude through the seed coat. Other seeds fail to develop and shrivel within flattened seed pods. The cut seed of deficient field peas reveal a sunken brown spot in the centre ('marsh spot').

Corrective measures

Soil applications   For most crops, including oats, wheat, lupins and soybeans, apply 20–50 kg/ha of manganese sulphate at sowing, banded with or near seed. Broadcasting and incorporation prior to sowing is less efficient than banding and requires higher rates (100–200 kg/ha). Topdressing applications after the crop is sown is the least effective method of applying manganese to the soil, because the soluble fertiliser manganese is rapidly changed to insoluble forms in the soil surface and is unavailable to roots. For the same reason, the residual effects of manganese treatments for subsequent crops are usually low, and annual re-treatment at rates of from half to the full initial rate are usually needed.

For an established crop, such as lucerne, topdress with 50 kg/ha of manganese sulphate or spray with 2 kg/100 L/ha.

Foliar sprays   Foliar sprays allow much lower rates per application (usually 4–5 kg manganese sulphate/100 L/ha), but more than one spray may be needed during the season to prevent symptoms reappearing. For sugar beet, rape and mustard, use 10 kg/100 L/ha. Timing is critical for crops such as sweet lupins where seed damage can be a serious problem. Spray lupin crops when the pods on the primary stem are about 3 cm long and flowering has nearly finished on the secondary stems.

1 **Oats** – a patch of pale stunted deficient plants within a mostly healthy crop. Severely affected plants suffer extensive leaf death and many die. Patchiness is common in manganese-deficient crops where areas of chlorotic plants coincide with sudden and obvious changes in the soil type. (J. Gartrell)

2

2 Oats – pale brown to grey necrotic lesions (called 'grey speck') develop between the veins. The symptoms first appear in mature leaves and progress into both old and young leaves. The lesions are often surrounded by a brownish purple 'halo'. Lesions can spread across the leaf, coalescing and causing it to collapse.

3 Oats – close up of necrotic lesions and faint interveinal chlorosis.

4 Maize – interveinal chlorosis of middle leaves is the first symptom of mild deficiency. When severe, older and younger leaves are also affected and white necrotic interveinal lesions develop within the chlorotic areas.

5 Sweet lupin – while still within the pod and immature, seeds of deficient plants split along the edge furthest from the point of attachment. The seed coat turns brown near the split. Other seeds shrivel and die within the pod. (J. Gartrell)

6 Sweet lupin – the cotyledons of split seed from deficient plants grow and protrude from the seed coat while still immature and within the seed pod. (J. Gartrell)

3

4

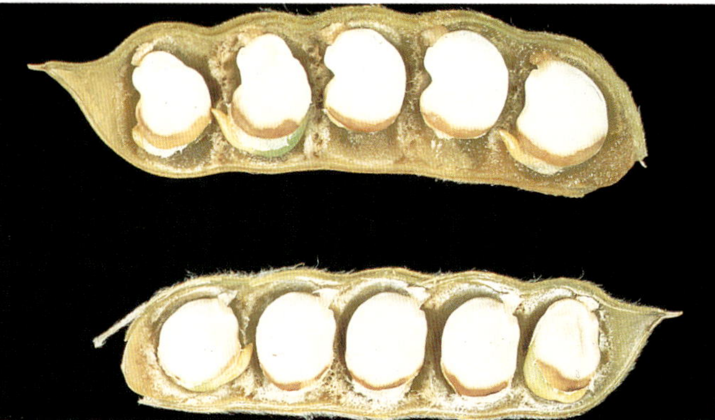

5

6

ZINC

Zinc deficiency most commonly affects crops grown in neutral or alkaline soils. In these soils, high pH and often high phosphate levels can hold zinc in forms not readily available to plants.

Zinc deficiency can be patchy and vary in severity from one season to another. Crops sown after a long fallow are more likely to develop acute deficiency, while cold wet soils will often produce severe symptoms in early planted summer crops during a cool spring. Land levelling for irrigation can accentuate zinc deficiency by exposing a more alkaline subsoil in the 'cut' areas.

Crops differ in their susceptibility to zinc deficiency. Sensitive crops include maize, linseed (flax) and some varieties of soybean. Sorghum and cotton are moderately sensitive, while sunflower and wheat are usually less affected.

Function   Zinc has an important role in the formation and activity of chlorophyll and in the functioning of several enzymes and the growth hormone auxin. Low levels of auxin in zinc-deficient crops causes the severe stunting of leaves and shoots which is so typical of this disorder.

Symptoms   A shortage of zinc is indicated most dramatically by a severe restriction in the growth and size of various tissues. Effects include small leaves, short internodes which give a compressed appearance to the plant (maize), or a complete cessation of terminal growth leading to a rosette of leaves and proliferation of side branches (linseed). Seed set can be greatly reduced and maturity delayed. Zinc deficiency also affects the colour of leaves, causing areas to become light green, yellow, white or bronze. The pale green or yellow areas may occur as a mottle between the veins (soybeans) or extend across veins as a broad bleached zone (maize). In cotton, the mottling between the veins can be bronze. Bronzing is also present in leaves of mature maize and soybean plants.

Maize or corn – Zinc deficiency in the early stages of growth (fourth to sixth leaf) is seen as bleached zones on each side of the midvein towards the lower half of the leaf. These pale zones may range from light green to creamy white in severe cases. The chlorotic patches in the leaf may die, causing the leaf to collapse. As the crop develops, interveinal yellowing (striping) becomes more common, and the chlorotic areas may become bronzed.

When the deficiency is severe, stem elongation can be restricted, giving the young corn plant a compressed 'trumpet' appearance.

Sorghum – Similar to those shown by maize, but often with a strong reddish purple coloration of the bleached or striped areas.

Linseed – Zinc deficiency causes extreme stunting. The growing tip ceases to elongate, producing a rosette of tiny leaves at the ends of shoots. Lower leaves shrivel and fall off, leaving bare sections of stems. Numerous side branches develop to replace the primary shoot, but these in turn stop growing. Leaves become mottled and develop greyish brown dead spots.

Soybeans – An affected area within a crop appears yellowish brown from a distance. Deficient plants are compact in their general growth. The leaves are smaller than normal, and the tissue between the veins becomes pale green, progressing to a yellow-bronze colour that may turn brown or grey and die. Symptoms are most pronounced in the lowest leaves, which often die and drop early. Few flowers form, and those pods that set are late maturing and often deformed. Seed yield is greatly reduced.

Cotton – The leaves become bronzed yellow, especially between the veins. Older leaves tend to be thickened and brittle and often cup upwards. Maturity is delayed.

Wheat – Affected wheat crops show patchy uneven growth. Leaves develop grey-green blotches on either side of the midvein at about the centre of the leaf. These patches die, causing leaf bending or collapse.

Sunflower – Sunflowers rarely show specific leaf symptoms, but growth and seed yield can be reduced in very deficient soils.

Medics and clover – The first sign of deficiency is usually seen in the mature leaves, which develop dull grey-green or bronze areas near the centre and base of each leaflet. Light brown dead spots may appear irregularly around the outer region of leaves. New leaves remain very small on shortened leaf stalks, producing a rosette of dwarfed, often cupped or malformed leaves at the centre of the plant. In the field, no specific symptoms may be noticed, even where yield is reduced by the deficiency.

Corrective measures   A preventive soil application of a zinc compound, such as zinc sulphate heptahydrate (23% Zn) or zinc oxide (60–80% Zn) broadcast and worked well into the soil prior to sowing, is better than corrective sprays applied to the crop after it develops symptoms. For pastures and winter cereals, such as wheat and oats, apply 0.8–2 kg Zn/ha (3.5–10 kg zinc sulphate heptahydrate or 1.0–2.5 kg zinc oxide); the lower rate is for sandy or acid soils and the higher rate for alkaline heavy textured (clay loam) soils. For high-demand crops such as maize or cotton, broadcast and work well into the soil 5–10 kg Zn/ha (20–40 kg zinc sulphate or 6–15 kg zinc oxide), depending on the severity of the deficiency and soil type. This application may remain effective for up to 5 years.

   If zinc deficiency is identified after the crop emerges, try a foliar spray of zinc sulphate (23% Zn) 0.5–1 kg/100 L/ha as soon as possible after the first signs of deficiency in the crop. The first spray should be no later than 3 weeks after emergence for a crop such as maize, with a repeat treatment 2 weeks later. Use the lower strength for more sensitive broad-leafed crops, such as sunflowers or soybeans, and the higher rate for maize, sorghum and wheat. Spray treatments may not fully prevent yield loss due to the initial set-back at the onset of deficiency.

1 Maize – a principal effect of zinc deficiency is stunting of the whole plant or of individual tissues. Stunted maize on the right, without zinc, compared with healthy maize on the left which was treated with zinc.

2 Maize – broad creamy white chlorotic bands near the base and mid-leaf areas and a compressed growth habit, are the most striking symptoms of zinc deficiency in maize.

3 Maize – stem elongation is restricted in this zinc-deficient maize seedling, resulting in a compressed plant with a tight whorl of small leaves. The youngest leaves can barely be seen emerging from the sheath of earlier formed leaves.

4 Maize – broad, bleached, white to creamy bands from the middle to the base of the leaf are the most easily recognisable symptom of zinc deficiency in maize.

5

6

7

8

5 Maize – a later stage often seen in mid growth stages of the crop, where the broad bleached chlorosis is replaced by 'striping' between the veins of later formed leaves.

6 Maize – bronzing of the chlorotic leaves often occurs late in the growth of the crop (silking stage). Varietal differences and the intensity of the deficiency also affect the various forms of symptom expression.

7 Maize – when zinc deficiency is severe, grain set is restricted, especially in crops like maize and soybeans. In maize, cobs may fail to develop at all, or set poorly like those at the top left of the photograph.

8 Sorghum – the lighter colour of this strip through the crop was the result of its earlier flowering, promoted by zinc treatment of the soil before planting. Delayed maturity is a common effect of zinc deficiency which can lead to additional yield losses if frosts come early.

9 Wheat – pale thin patches in a wheat crop affected by zinc deficiency. Patchiness can be the first sign of zinc deficiency in wheat, although other causes including waterlogging or nitrogen deficiency may be responsible.

10 Wheat – leaves of zinc-deficient wheat showing light brown irregular-shaped necrotic spots scattered along the leaves. These dead areas may join together, causing the leaf to collapse across the middle (leaf Zn = 10 ppm).

10

9

12

13

11

11 Wheat – symptoms are first seen in the middle leaves but, when the deficiency is severe, leaves of all ages are affected by chlorosis, necrosis and collapse. Leaves often develop a muddy green chlorosis and acquire an oily appearance (leaf at top right) as necrotic patches extend irregularly amongst chlorotic and dull green tissue (leaf Zn = 7 ppm).

12 Barrel medic – leaves of deficient medics and clovers are smaller than normal. New leaves may become very small, and are often cupped or misshapen. Petioles of the youngest leaves are shortened giving a rosette effect. Leaf colour is a dull green with large irregular patches of light grey burn and tiny purple spots (leaflet at top right).

13 Subterranean clover – leaves become light grey in colour with some purple pigmentation and a few scattered dark specks.

14 Subterranean clover – light grey-green chlorosis with brown necrotic spots irregularly scattered over the leaflets. The chlorotic pattern varies with variety, maturity, and severity of the deficiency.

15 Linseed – a rosette of tiny yellow leaves at the end of each shortened stem, multiple stems, and bare lower stem sections caused by the shedding of lower leaves. Greyish white dead spots appear on leaves.

16 Linseed – a close-up of the necrotic leaf spots and rosette at the end of a shoot.

17 Lupin – severely stunted growth. Small narrow distorted leaflets and shortened stems give a compressed appearance. Compare with the normal-sized healthy leaf at right.

14

15

16

17

18

19

20

18 Lupin – pale, distorted zinc-deficient leaves compared to normal leaf (far right).

19 Soybean – deficiency is first seen as a pale green mottle between the veins.

20 Soybean – as the deficiency intensifies, loss of colour between the veins becomes more complete. Brown or bronzed necrotic spots develop within the chlorotic areas and the leaflets tend to curl downward, die or fall off prematurely. Growth of the crop is stunted and maturity is delayed.

21 Cotton – small thickened leaves showing a faint mottled paleness between the veins. The pale areas often develop a bronze-green hue and leaves cup upwards. Healthy leaf is shown at bottom left.

21

*C*opper deficiency rarely restricts crop or pasture growth in New South Wales, Australia, although the health of grazing animals can be affected in some areas. However, wide tracts of land elsewhere in Australia and in other parts of the world are copper deficient. Coastal sands and soils derived from calcareous dune sands and marine limestones are commonly low in copper. In many reclaimed peaty soils, copper is strongly held by the high organic component, causing deficiency in crops such as wheat, oats and lucerne. Copper deficiency is less common in fine-textured soils (clay loams) derived from igneous rocks such as basalt.

Crops differ in their susceptibility to copper deficiency. Table 8 groups some crop and pasture plants on their observed responsiveness to copper. However, different cultivars within some species exhibit a wide range in their tolerance to low soil copper.

Table 8   Sensitivity of some crop and pasture plants to copper deficiency

Susceptible	Moderately susceptible	Moderately tolerant
Wheat	Oats	Rye
Rice	Barley	Triticale
Sudan grass	Maize	Pasture grasses
Lucerne	Sorghum	*Lotus* spp.
Barrel medic	Clovers	Soybeans
(*Medicago trunculatum*)	Linseed (flax)	Lupins
	Sunflower	Canola
	Tobacco	Field peas

Function   Copper is essential for photosynthesis, for the functioning of several enzymes, in flower and seed formation, for the production of lignin which gives physical strength to shoots and stems, and to vascular elements required for water movement in plants.

Symptoms   Although the effects of copper deficiency are varied, depending on the particular crop, some symptoms are common to many species. The youngest leaves are usually the worst affected, showing chlorosis, necrosis or distortion or a combination of these symptoms. Plant tops lack rigidity and may appear wilted while individual leaves roll, bend, or develop crinkling. Terminal dieback in the young shoots and sterile seedheads can dramatically reduce the yield of crops such as wheat and oats, while poor seed set will endanger the regeneration of annual pasture legumes such as subterranean clover. Copper-deficient crops usually look patchy, are stunted, and yield poorly. Specific symptoms of copper deficiency expressed by some crops are described below.

Wheat, oats and barley – The first signs are patches of poor growth within a healthy crop. Affected plants appear limp or wilted. Young leaves are pale and hang down. In time, they turn yellow and die at the tips, becoming light brown and rolling into a tube. When mild, the leaf base may remain green but, when severe, the whole leaf and sometimes the whole plant dies. Maturity is delayed and pollen production reduced, and there are increased numbers of sterile or partly filled heads. The sterile heads become white ('white head') **but frost, fungal or mouse damage can cause similar effects.**

Lucerne – Wilting of the young leaves followed by curving of the ends of the petioles causes the leaflets to be bent backwards. Leaf colour is normal but pale grey to white spots develop near the base or margins of leaflets. Death of these young leaves and the growing point follows.

Subterranean clover – Slightly stunted growth. Leaves are pale and erect, with younger leaves cupping. The unrolled margins develop a light brown necrosis while the remainder of the leaf is puckered. In the field, leaf symptoms may not be seen, but copper deficiency can seriously reduce flowering and seed set, resulting in poor re-establishment.

Maize – The crop is patchy and affected plants stunted and limp. Young leaves are pale yellow to white with interveinal and marginal chlorosis whereas old leaves can be a normal healthy green. The tips and margins of the youngest leaves burn, causing them to form a tube which fails to unroll and the whole leaf may die.

Flax (linseed) – Yellowing, distortion and death of upper leaves. Death of the growing point and rosetting at the top of the plant. Stems are stiff and thickened. Stunted growth and poor seed production.

Field peas – Plants are stunted, terminal growth wilts, pods are apparently normal but don't develop seeds.

Soybeans – No leaf symptoms or growth abnormalities are usually observed but no seed is set.

Copper deficiency is often difficult to diagnose from symptoms alone because its effects can be confused with other disorders (for example, frost damage in heading wheat). Also, yield or the regeneration of a pasture can be reduced without clear symptoms being produced, so leaf analysis may be required to confirm a deficiency.

Corrective measures   Applications of 5–50 kg/ha of copper sulphate (bluestone), depending on species and soil type, well incorporated into the soil can last for up to 10 years. As excess of copper may be phytotoxic; the lowest rate should be used on sandy soils low in organic matter, whereas heavier peaty or marl soils may need 20–50 kg.

A foliar spray of 2 kg/ha of copper oxychloride, at either high or low volume, applied at an early growth stage, is a safe, effective and relatively rainproof treatment for a growing crop. Copper sulphate used as a spray without neutralisation may cause leaf burn on sensitive crops, but for winter cereals, such as wheat and oats, a spray of 1–2 kg/ha has proved highly effective and caused negligible leaf scorch. Whereas an adequate soil application will correct a deficiency of copper for several years, foliar sprays must be repeated annually.

1 Wheat – (left) a normal copper-sufficient mature plant with tall straight stems and fully grain-filled heads. The centre plant displays mild deficiency symptoms including, delayed maturity (i.e. green heads and some leaves remaining green), white-tipped flag leaves and empty or partly filled ears. The more severely affected plant at the right, has very few emerged heads which are poorly filled, while other heads have failed to emerge. Stems are weak, leaves are twisted and many tillers have aborted. (J. Gartrell)

2 Wheat – a healthy grain-filled head (left) compared to a poorly filled deficient head (right). Spikelets near the tip are usually most affected, giving a typical 'rat-tailed' head. Compare this to frost damage where the empty spikelets can be found at any part of the head, depending on when frosting occurred in relation to head development. Darkening of the leaf sheaths stems and nodes is another sign of mild copper deficiency in wheat. (J. Gartrell)

Toxicities

Some elements, particularly chloride, sodium, manganese, boron and nitrogen, commonly cause injury when taken up in excess.

SODIUM AND CHLORIDE

When high concentrations of salt (sodium chloride) build up in the soil, most crop and pasture plants suffer a reduction in growth and productivity and usually develop toxicity symptoms. The problem occurs in many inland areas, but it is also found in coastal areas where the topsoil, subsoil or irrigation water is saline. There is a wide range of tolerance within commonly grown crop and pasture plants, from the relatively tolerant plants like barley, sugar beet and cotton to sensitive species like timothy grass and most of the clovers (Table 9). There are also differences between cultivars of many of these species.

Table 9 Sensitivity of crops and pasture plants to salt toxicity (listed in approximate order of increased tolerance down each column)

Sensitive	Moderately tolerant	Tolerant
Most species of *Trifolium* incl.:	Lucerne	Kikuyu grass (*Pennisetum*)
White clover	Strawberry clover	Wimmera ryegrass
Red clover	Sweet clover (*Melilotus*)	Couch grass (*Cynodon dactylon*)
Alsike clover	Birdsfoot trefoil (*Lotus*)	Barley grass
Subterranean clover	Siratro	Triticale
Timothy grass	Linseed (flax)	Cotton
Cocksfoot	Sunflower	Sugar beet
Dolichos	Rice	Cereal rye
Cowpea	Canola (rapeseed)	Salt bush (*Atriplex*)
Vetch	Soybean	Barley
Peanut	Safflower	
Maize	Cowpea	
	Paspalum	
	Brome grass	
	Phalaris	
	Panic	
	Sudan grass	
	Sorghum	
	Wheat	
	Oats	
	Tall fescue	
	Perennial ryegrass	
	Rhodes grass	

Burning of the leaf tip and yellowing or scorching of the margins are the most common symptoms of chloride or sodium toxicity. Leaf fall can be heavy and dieback may result. Older leaves usually show the symptoms first. Drought, potassium deficiency or fertiliser burn can produce similar symptoms, and leaf analysis may be needed to confirm the diagnosis.

Irrigation management and drainage are important means of combating salinity damage in irrigated crops and pastures. Improved practices can lower, or at least minimise, salt build up in the root zone. Test the quality of the irrigation water, avoiding sources which are saline. Water from some bores, tidal rivers or dams dug from saline shales can become salty during a long dry period, although they may have given satisfactory test values in wetter times. Salting in inland irrigation areas occurs when the salt-rich ground watertable moves upwards in the soil profile and nearer the surface. Over-watering or frequent cropping can cause the watertable to rise, then capillary action with alternate irrigation and drying cycles move salts to the surface.

In non-irrigated areas where salinity becomes a problem, more tolerant species or cultivars may be grown. In some cases, a drastic change in management, including the planting of strategically located tree lots, may be needed to lower the watertable and remove salt from the surface soil.

Cl1

Cl1 Lucerne – light brown to grey necrosis, initially of the tips and margins. Later, all of the distal parts of the leaflets become necrotic (leaf Cl = 3.2%).

Cl2

Cl3

Cl4

Cl5

Cl2 Lucerne – necrosis and cupping of the distal margins of leaflets. The symptoms occur first, and are most acute in the older leaves (leaf Cl = 3.5%). Chloride toxicity, potassium deficiency and manganese toxicity can exhibit similar symptoms, requiring a leaf analysis to confirm the diagnosis.

Cl3 White clover var. Ladino – light brown necrosis of the leaf margins and some interveinal tissue, with some inward curling, splitting and tearing of necrotic leaf edges. The whole leaf is a dull greyish green colour (leaf Cl = 3.5%).

Cl4 Maize – light grey-brown necrosis of the tips and margins in mature and older leaves of this seedling. Necrosis spreads back along the leaf margins from the tip, till most of the leaf is dead. Plants in the field have a drought-affected, dull greyish green appearance and growth is stunted (leaf Cl = 3.6%).

Cl5 Soybean – grey necrotic areas along the leaf margins of older leaves, spreading inward until only a narrow strip along the midrib remains alive and green. Finally the leaf dies and falls off (leaf Cl = 2.6%).

Cl6 Peanut – marginal yellowing (middle leaf), followed by brown necrosis and cupping of the leaf (right). Healthy leaf is shown at left.

Cl6

Na1 Subterranean clover – light grey necrotic semicircular spots caused by excessive sodium uptake. Necrosis first developed in interveinal tissue near the margin of leaflets but later coalesced into larger patches around the margin (leaf Na = 1.2%, Cl = 0.7%, K = 1.2%).

Na2 Subterranean clover – advanced burning of the leaf margins and interveinal areas caused by sodium toxicity.

Na3 Faba bean – dark brown to black necrosis of the leaf margin (leaf Na = 2.2%).

Na1

Na3

Na2

Na4

Na5

Na4 Lucerne – sodium–potassium imbalance and high chloride producing marginal leaf burn (leaf Na = 0.9%, Cl = 1.9%, K = 0.5%).

Na5 White clover var. Ladino – sodium–potassium imbalance caused this marginal chlorosis and necrosis. A high level of salt in the soil depressed potassium uptake causing combined symptoms of potassium deficiency and sodium toxicity (leaf Na = 1.3%, K = 0.5%).

Boron, an essential trace element which is often deficient, is also sometimes present in soils at levels which cause toxicity. With boron, there is a fine line between too little and too much. Boron toxicity has long been recognised as a disorder of sensitive fruit crops, such as citrus and grapes, in several inland irrigation areas. However, during a severe drought in 1982, boron toxicity was identified as the cause of symptoms and reduced yields in barley crops throughout large tracts of the wheatbelt of South Australia. Subsequently, boron toxicity has also been identified in wheat, annual medics and field peas.

Sometimes toxicity results from the over-use of boron fertiliser, but the usual cause in pastures and field crops is a naturally high boron content in many soils derived from boron-rich marine deposits. Soluble boron is easily leached, but in low rainfall areas, like those of extensive sections of the cereal growing sections of South Australia, western Victoria and parts of Western Australia, it remains in the root zone where it can be taken up in toxic amounts, especially in drought years. Boron toxicity can also arise when the boron content of irrigation water (commonly from bores) is too high.

The main symptom of boron toxicity is necrosis which affects the margin or tip of older leaves. Leaf venation influences the development of the pattern. In plants with parallel veins like grasses, barley or maize, necrosis begins from the leaf tip, while in broad-leafed plants, like peas or medics, yellowing and necrosis begins around the leaf margin. As the toxicity progresses, necrosis moves backwards from the tip to the mid-leaf area of cereals, or inwards (and sometimes interveinally) from the margins of peas, medics or other broad-leafed plants. Chlorosis may precede leaf burn.

Correction in irrigated crops where drainage is good, and there is ample high quality water, is by leaching with several heavy irrigations. Gypsum additions may help the leaching process by improving the soil's permeability. However, these measures are not appropriate where boron toxicity occurs in dryland pasture and field crops. Selection of more tolerant cultivars of barley, wheat, medic, field pea and other crops is probably the only way to minimise loss in these situations.

1 **Lucerne** – toxicity symptoms caused by excessive or careless application of boron fertiliser. Boron fertilisers must be used at recommended rates and spread evenly to avoid toxicity, as the range between too little and too much is narrow for this trace element.

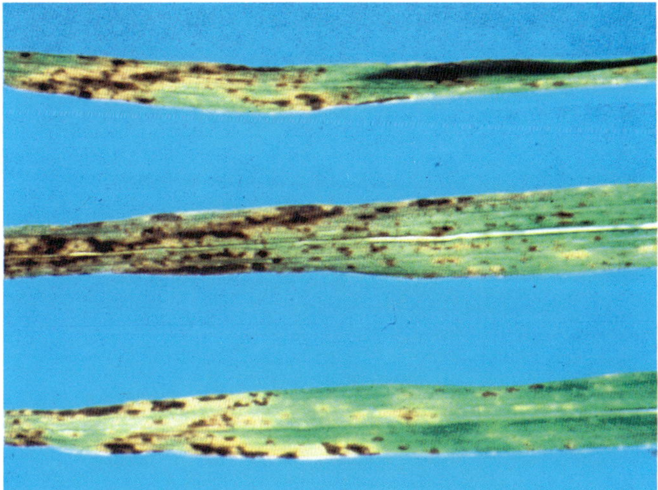

2 **Barley** – chlorosis and dark brown elongated spots develop along the blades of older leaves, especially near the tips. (J. Gartrell)

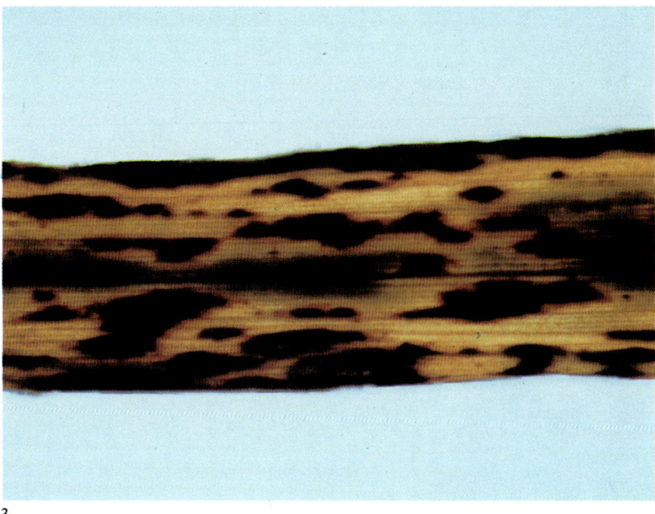

3 **Barley** – close-up section necrosis and areas of chlorosis with some green areas near the leaf midrib. (J. Gartrell)

*T*his is mostly a problem of acid soils, often made worse by periodic waterlogging or the use of acidifying ammonium fertilisers. Manganese toxicity occurs on a wide range of soil types. The red basaltic soils of temperate coastal and tablelands have long been recognised as being very high in available manganese, but the problem also occurs in many other soils including some sandy loam soils in widely scattered localities (Table 10). Climatic extremes experienced in the predominantly winter rainfall climate of southern tablelands of New South Wales – waterlogging in winter and heating of bare fallowed soils in summer – often exacerbate the acidity problem, raising available manganese in soil to toxic levels. When peak levels coincide with the establishment of susceptible crop or

Table 10   Occurrences of manganese toxicity in some New South Wales crops identified in the period 1957–88

Crop	Occurrences	Range in leaf Mn (ppm)	Principal locations in NSW, Australia
Lucerne	39	460–2890	Southern Tablelands, Bathurst, Goulburn, Albury, Wagga
Rape	13	508–3650	Central and Southern Tablelands, Crookwell, Oberon, Tumut
Turnip	5	1030–2480	Central and Southern Tablelands
Lupin	8	2400–7500	Western Slopes
Soybean	3	1000–2525	North Coast (Grafton), Central Coast (Camden)
Sunflower	2	3860–8225	Southern Tablelands (Crookwell) and Central Coast (Camden)
Field pea	1	4108	Southern Slopes (Finley)
Faba bean	2	960–1000	South Western Slopes
Tobacco	2	1200–1300	Northern Slopes
Subterranean clover	3	1272–2090	Southern Tablelands
Red clover	1	2000	Northern Tablelands

pasture plants such as canola (rapeseed) or lucerne, damage to seedlings can be severe.

Crop and pasture species show a wide range of sensitivity or tolerance to high manganese in the soil. Plants like lucerne, rape and lespedeza are very sensitive to high soil manganese, while oats, cotton and most pasture grasses are relatively tolerant (Table 11). In some species, such as soybean, the range of tolerance between the cultivars can be very wide.

A common symptom is yellowing of the margin of the older leaves (lucerne) often followed by cupping (canola). In some plants small dark or reddish brown necrotic specks develop over the leaf surface and along the veins and petioles of older leaves (soybeans).

The principal means of control is the correction of soil acidity but soil drainage should also be examined. Apply agricultural lime at the rate of 1 tonne/ha or more, depending on soil pH and type. Heavier textured soils will need more lime than sandy soil to lift the pH to a desirable 6.0–6.5 (1:5 water).

Table 11   Sensitivity of some crops and pasture plants to excess manganese and aluminium

Sensitivity	Manganese	Aluminium
Sensitive	Lucerne Annual medics Canola (rapeseed) *Melilotus* spp. *Glycine* Soybeans[1]	Lucerne Annual medics Barley Phalaris seedlings Canola Wheat (e.g. Egret)[1] Sunflower Red clover
Moderately tolerant	Wheat Narrow-leaf lupin Cowpea Soybeans[1] *Crotalaria* *Dolichos lablab*	Maize Oats[1] White clover Subterranean clover White lupins Turnip Tall fescue Townsville stylo Triticales[1] *Glycine*
Tolerant	Barley Maize Oats Rice Phalaris Soybeans[1] White lupin Subterranean clover White clover Cotton Sunflower Most grasses	Triticales[1] Narrow-leaf lupins Cocksfoot Cereal rye Kikuyu Paspalum *Setaria* Siratro *Dolichos lablab*

[1] The range of tolerances between cultivars can be large.

Soil and crop management practices may also need changing. Chronic poor drainage, or even temporary waterlogging following irrigation or the first autumn rains, can accentuate the effects of soil acidity, causing high manganese levels to peak soon after seedlings germinate, stunting or killing them. Fallowing during the heat of summer can also elevate available soil manganese levels, seriously damaging young plants of autumn-sown rape or lucerne. Alternate ground preparation and sowing strategies may be needed to avoid such toxic peaks coinciding with the most susceptible stage of a crop, namely seedling establishment.

1

1 **Lucerne** – yellowing of the margin of leaflets of older leaves. Chlorosis is mostly confined to the distal (tip) half of the leaflet, with the basal part of the leaf and its margin remaining green.

2 **Lucerne** – in old leaves, the marginal chlorosis can resemble potassium deficiency symptoms being a creamy yellow rather than golden (cf. previous photo), and there can be some speckling (leaf far right). Unlike potassium deficiency, chlorosis due to manganese toxicity is largely confined to the distal half of the leaflets (leaf Mn = 1420 ppm, K = 3.5%).

2

3

4

5

3 Subterranean clover – chlorosis of the margin of the distal half of leaflets of mature leaves. The leaflet edge finally becomes necrotic (leaf on right).

4 Subterranean clover – later stages of toxicity symptoms in old leaves, with chlorosis being replaced by necrosis (left leaf) and grey necrotic tissue covering much of the outer parts of the leaflets (right leaf).

5 Canola (rapeseed) – a patch of stunted crop with yellow cupped leaves (leaf Mn = 2280 ppm). Patches of manganese toxicity in a crop will often coincide with low wet areas or changes in the soil colour, texture, or pH. Field observations made of the crop are as important as leaf symptoms or chemical analysis when making a diagnosis.

6 Canola – yellowing of the leaf margins (centre and right) and leaf cupping (left) (leaf Mn = 3425 ppm).

6

7 Canola – chlorosis develops mainly around the margins of the distal part of leaves of all ages but some chlorosis can extend inward between the veins (leaf Mn = 3560 ppm).

8 Mustard – this crop is sensitive to excess manganese like many other members of the crucifer family which includes canola. Chlorosis of the leaf margin especially towards the leaflet tips. Light brown necrotic spots with darker brown borders develop within the chlorotic areas.

9 Lupin – rusty red leaf coloration (left), compared to normal leaf (right) (leaf Mn = 3600 ppm).

7

8

9

10

11

12

13

10 Lupin – late stages of severe toxicity displaying reddish brown necrotic spots mostly towards the edge of leaflets causing some leaf puckering and downward curling of the margin. Tissue along the leaf midribs still retains its green colour.

11 Field pea – chlorosis quickly followed by whitish necrosis of the distal margins of leaflets and stipules.

12 Soybean – chlorosis and grey necrotic areas of the margins and interveinal areas of young leaves. Some puckering ('crinkle') of the affected leaflets is also evident. Note the tiny rusty brown necrotic specks within the interveinal chlorotic areas. Similar tiny brown specks appear along the main veins on the undersides of leaves (leaf Mn = 4200 ppm).

13 Sunflower – this crop is relatively tolerant of high manganese but in extreme conditions plants become stunted and mature leaves chlorotic, and then cupped, with brown necrosis appearing between the veins and around the margins (leaf Mn = 8225 ppm).

14 Sunflower – progression of toxic symptoms. Healthy leaf (left), chlorosis from the leaf margin extending interveinally (centre), followed by leaf necrosis of interveinal tissue and the margins, causing inward rolling of the leaf (right). The petioles of affected leaves are marked by tiny dark specks (localised deposits of manganese dioxide).

15 Tobacco – leaves develop a dull rusty brown pale colour (leaf Mn = 1294 ppm).

16 Tobacco – when magnified the brown leaf colour is seen to consist of many tiny brown spots in the interveinal areas. These are crystals of manganese dioxide, deposited in older leaves usually near vein endings.

14

15

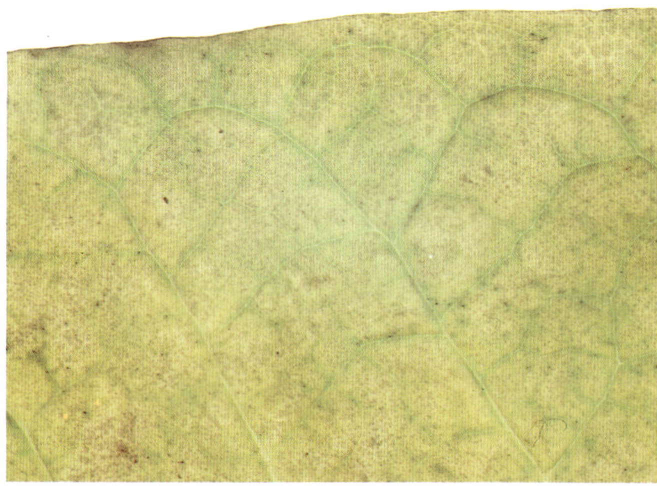

16

ALUMINIUM

The other major toxicity problem affecting plants growing in acid soils is aluminium toxicity. Although aluminium is a principal component of many soil minerals, it is only a problem to plants when it becomes soluble in very acid soils (usually pH 5 or less 1:5 calcium chloride).

In general, plants respond to aluminium quite differently than they do to manganese. Whereas manganese is taken up readily and moves to the plant tops causing symptoms of high manganese levels in the leaves, absorbed aluminium mostly accumulates in the roots. Consequently, plants suffering from aluminium toxicity normally exhibit few symptoms in their tops, apart from poor growth, but roots may be very stunted, thickened and stubby with darkened, dead spots. Aluminium accumulated in roots often restricts the uptake of other nutrients, particularly phosphorus and sometimes calcium, magnesium or potassium. Because of this, a superficial examination of plant symptoms can suggest phosphate deficiency and leaf analysis for aluminium is usually not a reliable guide to this toxicity. A soil analysis which measures the percentage aluminium saturation of the soil cation exchange provides the best guide to aluminium toxicity. A soil pH test alone is not sufficient.

There is a wide range of tolerance to aluminium among pasture and crop species and cultivars. Lucerne and barley are quite sensitive, most cultivars of subterranean clover and oats are moderately tolerant, and narrow-leaf lupin and cereal rye are some of the most tolerant of crop plants. A tolerance of high manganese does not always imply that a species or variety is also tolerant of aluminium (see Table 11). For example, phalaris and the Egret variety of wheat are both tolerant of high manganese but quite sensitive to aluminium.

Since aluminium toxicity is a problem of acid soils, liming is the usual method of correction. Apply agricultural lime at the rate of 1 tonne/ha or more, depending on soil pH and type. Where economic considerations limit the use of the high rates of lime often needed to raise the soil pH to a safe level (pH 5.5–6.0 1:5 $CaCl_2$), selection of more aluminium-tolerant species (see Table 11) may enable cropping or pasture production to continue. However, this may only provide breathing space, since most forms of cropping and pasture production will continue to cause the soil pH to fall unless some lime is used.

1 Lucerne – plants of this sensitive species are stunted and dying in the plot (left foreground) which received only 220 kg of lime/ha. This rate was insufficient to effectively reduce the high exchangeable aluminium level in this acid soil to a level sufficiently low for satisfactory growth of lucerne. Compare the better growth of lucerne (right) receiving 2200 kg of lime/ha and 4400 kg/ha (top left). All plots received molybdenum and were innoculated with Rhizobium.

2 Lucerne – application of 4400 kg/ha agricultural lime (centre plot) reduced exchangeable aluminium to an acceptably low level and allowed lucerne to establish in this acid soil. Lucerne sown in unlimed plots on either side failed due to aluminium toxicity.

3 Wheat – stunted growth, particularly of the roots, is a principal effect of high levels of exchangeable aluminium in the soil. Note the few short stubby thickened roots of this stunted plant.

1

2

3

NITROGEN *(FERTILISER BURN)*

Fertiliser burn to leaves and roots is usually caused by fertiliser mistakenly applied at too heavy a rate or too close to roots, or by direct contact of fertiliser with leaves during topdressing. Excessive or careless application of any nutrient can cause injury, but symptoms observed soon after the application of a fertiliser are most often due to a sudden uptake of soluble nitrogen salts. A quick tissue test for nitrate in the leaf tissue showing the burn, or immediately adjacent to it, will usually reveal high levels of nitrate ($>10\,000$ ppm). Normally, the tissues around the edge of healthy leaves are relatively low in nitrate (<1000 ppm).

Young seedlings are particularly at risk, but all plants can burn if a fertiliser application is heavy handed or uneven. Symptoms of fertiliser burn develop quickly, often within a few days of an application. Leaf curl, defoliation, death and blackening of the shoot tip are the most obvious effects. Young recently matured leaves are usually the first to show injury. Symptoms include irregularly distributed patches of translucent dark green tissue which later become brown and dead. In other cases, the leaves develop a narrow band of chlorosis and then scorch along the leaf margin. Leaves are often dark green and may droop and curl downward. Wilting when soil moisture is adequate is common.

Although root damage may not be noticed, it is usually present. Injured roots are dull brown instead of light in colour and are susceptible to fungal attack. Damage is often irreversible but immediate leaching of soil with a generous irrigation will assist plant recovery.

1 Cotton – light straw-coloured irregularly shaped burn of the outer parts of the leaf caused by excess fertiliser applied too close to young plants.

NICKEL

1

2

1 Oats – the stunted pale patch of oats in this crop is the result of naturally toxic levels of nickel in this soil derived from a serpentine rock rich in nickel. Toxicities of nickel or other heavy metal can also occur in crop and pasture plants used to revegetate areas of old mining wastes.

2 Oats – nickel toxicity produces characteristic symptoms comprising transverse bands of light brown to white chlorotic and necrotic tissue. Oats is a useful indicator plant of nickel toxicity because of the unique symptom pattern it produces and its relative sensitivity to nickel.

Some non-nutritional symptoms

A number of non-nutritional disorders or stresses cause symptoms in leaves or other tissues that are mistaken for nutrient disorders. Close examination of the symptom pattern, however, will often reveal that the resemblance is superficial and that the pattern is basically different to the symptom caused by a nutrient deficiency or toxicity.

Non-nutritional symptoms can arise from many environmental stresses, including frost, heat, drought, or wind, from physiological effects resulting from root injury or waterlogging, and from spray burns, herbicides, genetic abnormalities, pests and diseases.

Vein chlorosis   The terms 'vein chlorosis', 'vein clearing' or 'yellow vein' are used to describe the loss of green colour from the midrib and major veins. The interveinal tissue remains green, though it may be paler than normal. This pattern is contrary to common nutritional patterns where colour loss begins between the veins. Vein chlorosis normally indicates root injury which can be caused by a number of factors including disease, waterlogging, or exposure to certain herbicides.

Herbicides   Herbicides cause a variety of symptoms in leaves when taken up either through the roots or by direct contact of the spray with leaves.

Herbicide symptoms include stunting, distortion of leaves or shoots, vein clearing, and a variety of chlorotic leaf patterns. Chlorosis caused by a herbicide usually differs from typical nutrient deficiency symptoms as pattern symmetry is usually absent; the chlorotic areas are unrelated to leaf venation, or chlorosis begins in the veins and moves outward from them. The colour of the chlorosis is often an almost artificially bright hue of yellow, orange, cream or white.

Spray burn   Sprays of trace elements and other nutrients or pesticides can injure leaves, if they are applied incorrectly – at the wrong strength or time, during heatwave conditions, or to plants which are moisture stressed. Injury can occur if materials are not correctly mixed or kept agitated in the spray vat. Where several materials are used together in the one spray, they must be compatible (i.e. they won't react) and, therefore, not produce a residue which could damage leaves.

Spray injuries will vary with the chemical used, the conditions at spraying, the spray equipment, the type and condition of the plant, and the maturity of the tissue at the time of application. A clue to this type of damage can often be found in the injury pattern which frequently follows the shape and distribution of the spray droplets. Another indication is the suddenness with which symptoms develop. Spray burn may appear within days of the spray application with many plants developing the symptoms at the same time. Nutrient disorders tend to show up to different degrees throughout the crop; the symptoms first appearing in only a few patches, then gradually becoming more evident.

Symptoms caused by adverse environmental factors such as frost, hail, heat or wind, may be seen within hours or at most a few days after an abnormal weather event. The sudden appearance of symptoms can be an important clue to the nature of a problem in a crop.

Frost   Leaves or other plant parts develop dry necrotic areas on the exposed parts which may at first appear grey-green in colour. Leaves first wilt and later develop dry burnt patches. Midribs and petioles can become brittle and develop splits or cracks. Young plants are most at risk with the growing point often being killed. Frost damage is usually worse in the low lying parts of the field. In cereals and grasses, chlorotic bands may develop across the leaf blade as a response to short periods of cold weather.

Drought and heat   If the stress is only moderate and for short duration, temporary wilting may be all that happens, leaving no permanent symptom. However, severe moisture stress will cause irregular shaped burning of the tips or outer parts of leaves which may then yellow and die. Sunscald, caused by sudden heatwave conditions, can produce similar effects on leaves even when soil moisture is adequate. The distribution of healthy and damaged tissue is often related to the presence or absence of shade from protecting leaves.

Hail damage   Hail causes perforated and torn leaves or even complete defoliation.

Wind injury   Wind effects include leaf tatter, distortion, abrasion from sand blasting, and girdling where the crown of the plant is rubbed against the soil. Windburn sometimes causes a silvery sheen on the wind-exposed surface of leaves.

Genetic abnormalities   These tend to be found as a random scatter of affected plants throughout a crop rather than in the patches that are usual with nutritional disorders. Isolated plants may be stunted and distorted or show irregular yellow patterns in leaves.

1

2

3

4

1 Maize – wilted plants (left) resulting from faulty irrigation practices. Leaves roll inward, exposing the under surfaces of leaves which can develop sunburnt patches. (A. Dale)

2 Maize – severe frost injury to a young crop planted too early in the spring. The upper, more exposed leaves are most affected. This damage becomes apparent in many plants simultaneously, within a few days of a heavy frost. (A. Dale)

3 Wheat – late frosts occurring at ear formation caused the empty spikelets in the two heads on the left. Frost injury to heads can be confused with the effects of copper deficiency. Copper deficiency usually affects the tip of the head more than the lower spikelets, producing a 'rat-tailed' head, while frost injury can occur along any part of the head. (H. Marcellos)

4 Wheat – cold injury, the result of a succession of very cold nights during emergence, causes bands of chlorosis regularly spaced along young leaves. This effect is distinct from frost injury which causes death of leaf tissue.

5

6

8

7

5 White clover – herbicide damage causing irregularly shaped whitish patches of chlorosis. The bottom leaf has had all its chlorophyll destroyed by the herbicide apart from the main vein and part of the margin.

6 Tobacco – bright yellow interveinal chlorosis caused by the incorrect use of a herbicide. Chlorosis in this case is symmetrical and interveinal, but herbicide-induced chlorosis is often irregular in pattern and the colour of the chlorosis is white or an almost unnatural bright yellow or orange hue.

7 Wheat – herbicide uptake has caused the plant on the right to become almost completely white. The slightly affected plant on the left has developed one white patch near the base of its second leaf and mild chlorosis in other parts.

8 Lupin – necrotic spots and pale grey-green chlorosis over most of the leaf except for tissue along the midrib caused by the herbicide simazine. (G. Stovold)

10

9

11

9 Lucerne – irregularly patterned bright yellow chlorosis of leaves of a plant affected by the alfalfa mosaic virus. (G. Stovold)

10 White clover – a variety of irregular chlorotic leaf patterns caused by a virus infection.

11 Canola (rapeseed) – scattered yellow spots caused by the fungus downy mildew. Notice in the leaf on the left, that each small brown necrotic spot is surrounded by a pale chlorotic 'halo'. The leaf on the right shows a more advanced stage of this condition. (G. Stovold)

12 Maize – leaf blight disease caused by a fungus infection. (G. Stovold)

12

13 Maize – interveinal elongated chlorosis and necrosis caused by virus infection can be confused with some symptoms of boron deficiency (14) or a late season symptom of zinc deficiency (15).

14 Maize – these boron deficiency symptoms can sometimes be difficult to distinguish from those of virus infection.

15 Maize – late season zinc deficiency symptoms can sometimes resemble those caused by virus infection.

13

14

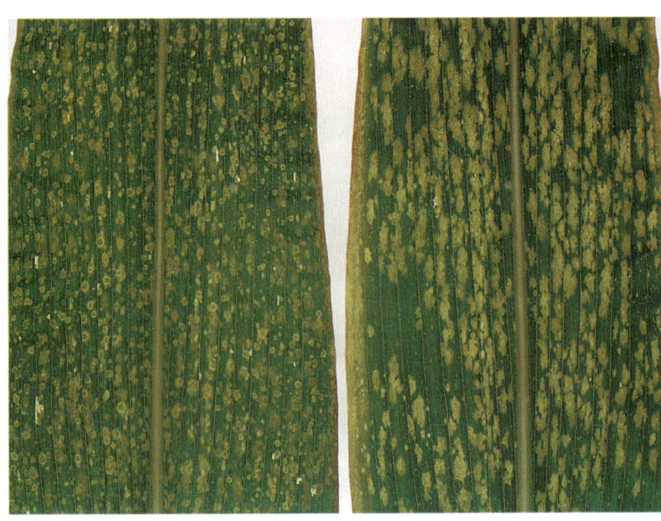

15

Appendix A

Leaf analysis standards for some important crops and pasture plants based on diagnostic and research analyses.

Canola (rapeseed)

(*Brassica napus* and *B. campestris*)

Plant part   Recently matured leaf
Growth stage   Pre-flowering

Nutrient	Deficient	Low	Normal	High	Excess or toxic
Nitrogen % (N)	0.8[D]–2.7[D]	2.8–3.2	3.5–5.5	5.6–6.5	
Phosphorus % (P)	0.09[D]–0.20	0.24–0.29	0.35–0.60	0.65–0.70	
Potassium % (K)	<1.6	1.8–1.9	2.8–5.5	6.5–8.0	
Sulphur % (S)		0.25–0.50	0.6–1.0		
Calcium % (Ca)		0.8–1.2	1.4–3.0		
Magnesium % (Mg)	0.14[D]	0.16–0.19	0.21–0.65		
Sodium % (Na)			0.02–0.5	0.7–1.1	1.7[T]
Chloride % (Cl)			0.4–1.6	2.0	
Copper ppm (Cu)		2–3	4–25		
Zinc ppm (Zn)	<12?	12–17	21–55		
Manganese ppm (Mn)			30–250	300–400	530[T]–3650[T]
Boron ppm (B)	6[D]–13	17–20	22–50		
Molybdenum ppm (Mo)			0.28–0.55	1.6	

[D] Showing deficiency symptoms.
[T] Showing toxicity symptoms.

Lucerne
(Medicago sativa)

Plant part Midstem leaves
Growth stage Early flowering (10%)

Nutrient	Deficient	Low	Normal	High	Excess or toxic
Nitrogen % (N)	<3.5	3.5–4.0	4.2–5.5	5.6–6.0	
Phosphorus % (P)	<0.20	0.20–0.23	0.25–0.50	0.60	
Potassium % (K)	<1.0	1.0–1.4	1.5–3.5	3.8–4.5	
Sulphur % (S)	<0.20	0.20–0.22	0.23–0.60	>0.60	
Calcium % (Ca)	<1.0	1.0–1.4	1.5–3.2	>3.5	
Magnesium % (Mg)	<0.18	0.18–0.22	0.25–0.50	0.65–0.85	
Sodium % (Na)			<0.3	0.5–0.6	0.8^T–1.5^T
Chloride % (Cl)			<1.5	1.5–1.7	1.8^T–2.8^T
Copper ppm (Cu)	<4	4	5–30	40–50	
Zinc ppm (Zn)		15–18	20–50	60–90	
Manganese ppm (Mn)		20–24	25–190	200–350	450^T–2900^T
Boron ppm (B)	<18	18–20	21–50	100–200	>200
Molybdenum ppm (Mo)	0.1–0.3[1]	0.4	0.5–5.0	6–10	>10

[1] Not a reliable guide alone unless associated with low total nitrogen.
[T] Showing toxicity symptoms.

Maize (corn)
(Zea mays)

Plant part Ear leaf
Growth stage Tasseling–initial silk

Nutrient	Deficient	Low	Normal	High	Excess or toxic
Nitrogen % (N)	<2.4	2.4–2.6	2.7–3.5	3.6–4.0	>4.0
Phosphorus % (P)	<0.16	0.16–0.24	0.25–0.45	0.46–0.80	>0.80
Potassium % (K)	<1.2	1.2–1.6	1.7–2.7	2.8–4.0	>4.0
Sulphur % (S)	<0.12	0.12–0.16	0.2–0.5	0.6–0.8	>0.8
Calcium % (Ca)	<0.1	0.1–0.2	0.2–0.5	0.6–0.9	>0.9
Magnesium % (Mg)	<0.12	0.12–0.15	0.16–0.60	0.61–0.85	>0.85
Sodium % (Na)			<0.3	0.4–0.5	>0.5
Chloride % (Cl)			<1.0	1.0–1.4	>1.8
Copper[1] ppm (Cu)	<2	2–5	6.0–20	21–50	>50
Zinc[1] ppm (Zn)	<17	17–18	18–60	70–150	>150
Manganese[1] ppm (Mn)	<15	15–19	20–200	200–300	3000[T]
Iron[2] ppm (Fe)		10–20	30–200		
Boron ppm (B)	<2	2–4	5–30	30–60	>60
Molybdenum ppm (Mo)		0.1–0.2	0.2–3.0	>3.0	

[1] Values for copper, zinc or manganese in leaves sprayed with fungicides or nutrient sprays containing trace elements cannot give a reliable guide to nutritional status even in washed leaves.
[2] Leaf analysis is not a reliable guide to iron deficiency because of surface contamination with dirt, or immobility of iron within the plant, or the presence of physiological inactive iron within tissues.
[T] Showing toxicity symptoms.

Rice

(Oryza sativa)

Plant part Mid tillering (Feekes 3–5)
Growth stage Y leaf i.e. most recently expanded leaf

Nutrient	Deficient	Low	Normal	High	Excess or toxic
Nitrogen % (N)	<2.5	2.5–2.9	3.0–4.5	5.5	
Phosphorus % (P)	<0.10	0.10–0.12	0.20–0.50	>0.55	
Potassium % (K)	<1.2	1.2–1.4	1.5–3.5	>3.5	
Calcium % (Ca)			0.10–0.30		
Magnesium % (Mg)		<0.12	0.14–0.25		
Sodium % (Na)			0.01–0.20	0.4–0.5?	
Chloride % (Cl)			0.1–0.6	1.9–2.2	2.5?
Copper ppm (Cu)		3–5	7–15		
Zinc ppm (Zn)	<15	15–18	20–60		
Manganese ppm (Mn)			40–500	600–1000	>5000
Boron ppm (B)		<5	5–15		
Molybdenum ppm (Mo)			0.2–5.0		

Grain sorghum
(*Sorghum vulgare* and *S. bicolor*)

Plant part   3rd leaf below the head
Growth stage   Just prior to flowering and early flowering

Nutrient	Deficient	Low	Normal	High	Excess or toxic
Nitrogen % (N)	1.0[D]–2.1[D]	2.5–2.8	3.0–4.2	5.0–6.5	
Phosphorus % (P)	0.11[D]–0.15[D]	0.17–0.23	0.25–0.45	0.5–0.7	0.81[1]
Potassium % (K)	0.4[D]–0.7[D]	0.9–1.3	1.6–3.3	3.5–5.0	
Calcium % (Ca)		0.12–0.20	0.30–0.60	0.8–1.0	1.2–1.5
Magnesium % (Mg)	0.11[D]–0.13[D]	0.15–0.17	0.20–0.50	0.6–0.8	0.9–1.2
Sodium % (Na)			0.01–0.20		
Chloride % (Cl)			0.10–1.5	1.8	2.2[T]
Copper ppm (Cu)		1	2–15		
Zinc ppm (Zn)	8[D]–11[D]	12–18	20–60	70–80	
Manganese ppm (Mn)	<8	8–10	15–350		
Boron ppm (B)			2–25	35–60	
Molybdenum ppm (Mo)		0.11	0.3–3.0		

[1] Imbalance, from crop showing zinc deficiency symptoms.
[D] Showing deficiency symptoms.
[T] Showing toxicity symptoms.

Soybean

(Glycine max)

Plant part   Upper most fully developed leaf
Growth stage   Commencement of flowering

Nutrient	Deficient	Low	Normal	High	Excess or toxic
Nitrogen % (N)	<3.2	3.2–4.0	4.2–5.5	6.0–7.0	
Phosphorus % (P)	<0.15	0.25	0.30–0.55	0.6–0.7	0.8–2.4[1]
Potassium % (K)	<0.8	1.0–1.5	1.7–3.2	4.0	
Sulphur % (S)		0.15	0.2–0.4		
Calcium % (Ca)		0.21–0.35	0.40–2.0	2.5–3.0	
Magnesium % (Mg)		0.15–0.19	0.25–0.80	1.0–1.2	
Sodium % (Na)			<0.1	0.3	>0.5
Chloride % (Cl)			<1.2	1.5–2.0	2.6–5.0
Copper ppm (Cu)		4–5	6–30	31–50	
Zinc ppm (Zn)	6–14	15–20	25–80		
Manganese ppm (Mn)	<15	15–25	30–100	150–250	750–1000
Boron ppm (B)		15–20	21–60	80–200	
Molybdenum ppm (Mo)	0.12	0.15–0.40	0.5–5.5	10	

[1] Nutrient imbalance associated with zinc deficient crop.

Subterranean clover and white clover
(Trifolium subterraneum and T. repens)

Plant part   Green leaves with petioles[1]
Growth stage   Pre-flowering to commencement of flowering

Nutrient	Deficient	Low	Normal	High	Excess or toxic
Nitrogen % (N)		3.0–3.2	3.3–5.5		
Phosphorus % (P)	<0.20	0.22–0.23	0.25–0.50	0.6–0.7	>0.7
Potassium % (K)	<0.8	0.8–1.0	1.1–2.5		
Sulphur % (S)	<0.22	0.22–0.23	0.25–0.40		
Calcium % (Ca)		<0.8	0.8–2.5		
Magnesium % (Mg)	<0.12	0.12–0.14	0.15–0.50	0.7	
Sodium % (Na)			<0.5	0.5–0.7	0.8[T]–2.4[T]
Chloride % (Cl)			<1.6	1.7–1.9	2.2[T]–6.8[T]
Copper ppm (Cu)	<3.0	3–4	5–30		
Zinc ppm (Zn)	<12	12–14	15–50	>50	
Manganese ppm (Mn)	<20	20–24	25–300	400–600	>800
Boron ppm (B)	<20	20–24	25–100		
Molybdenum ppm (Mo)	<0.4[2]	0.4	0.5–10	>10	

[1] For tall rank growth reduce petioles to 7–10 cm.
[2] Not a reliable guide alone unless associated with low total nitrogen.
[T] Showing toxicity symptoms.

Sunflower
(Helianthus annuus)

Plant part   3rd–4th leaf below the flower bud
Growth stage   Florets about to emerge

Nutrient	Deficient	Low	Normal	High	Excess or toxic
Nitrogen % (N)	<3.0	3.0–3.5	3.6–5.0	5.1–6.5	
Phosphorus % (P)	<0.25	0.25–0.29	0.30–0.60	0.7–0.9	
Potassium % (K)	<1.2	1.5–2.5	3.5–6.0	6.5–8.0	
Sulphur % (S)			0.3–0.5		
Calcium % (Ca)		0.6–1.4	1.5–3.5	3.8–4.8	
Magnesium % (Mg)	<0.11	0.11–0.18	0.3–1.2	1.5–2.4	
Sodium % (Na)			<0.3	0.5–0.6	>1.2
Chloride % (Cl)			<2.0	2.1–2.5	>2.5?
Copper ppm (Cu)		3–4	5–50	60–70	
Zinc ppm (Zn)	<15	15–18	30–80	90–200	
Manganese ppm (Mn)	<15	15–20	40–850	1500–2000	>3000?
Boron ppm (B)	<20	25–30	35–90	100–200	
Molybdenum ppm (Mo)	<0.25	0.25	0.3–0.6		

Wheat, barley and oats
(Triticum aestivum, Hordeum vulgare and *Avena sativa)*

Plant part   Recently matured leaf
Growth stage   Mid to late tillering

Nutrient	Deficient	Low	Normal	High	Excess or toxic
Nitrogen % (N)	<3.4	3.4	3.5–5.4	5.5–6.5	>6.5
Phosphorus % (P)	<0.24	0.24–0.29	0.30–0.50	0.6–0.7	>0.7
Potassium % (K)	<1.5	1.5–2.3	2.4–4.0	4.1–5.5	>6.0
Sulphur % (S)		<0.15	0.15–0.40	>0.40	
Calcium % (Ca)		<0.18	0.21–0.40	0.6–0.7	
Magnesium % (Mg)	<0.11	0.11–0.12	0.13–0.30		
Sodium % (Na)			<0.5	0.6–0.7	>0.8
Chloride % (Cl)			<2.0	2.0–2.8	>3.0
Copper ppm (Cu)		2–4	5–50		
Zinc ppm (Zn)	<14	14	15–70		
Manganese ppm (Mn)	<12	12–24	25–300	400–600	700[T]
Boron ppm (B)	<2	2–4	5–10	11–20	30–100
Molybdenum ppm (Mo)	<0.05	0.05–0.09	0.10–0.50	0.6–0.7	

[T] Showing toxicity symptoms.

Appendix B

The standards for the following crop and pasture plants are based largely on diagnostic leaf analyses. Sampling times are defined as a stage of growth and are appropriate for both Southern and Northern Hemispheres.

Barrel medic and burr medic
(Medicago trunculata and M. denticulata)

Plant part   Mature leaves
Growth stage   Pre-flowering

Nutrient	Deficient	Below normal	Normal	Above normal	Excess or toxic
Nitrogen % (N)	1.8[D]–2.7[D]	3.0–3.2	3.5–5.5		
Phosphorus % (P)		0.21	0.26–0.45		
Potassium % (K)			1.5–4.0		
Sulphur % (S)	0.11[D]	0.16–0.20	0.25–0.60		
Calcium % (Ca)			0.8–1.2		
Magnesium % (Mg)		0.17	0.20–0.55		
Sodium % (Na)			<0.10–0.35	0.45	1.0[T]
Chloride % (Cl)			0.10–1.4	1.8	
Copper ppm (Cu)	<4	4	6–25		
Zinc ppm (Zn)		17	25–60		
Manganese ppm (Mn)			35–250		
Boron ppm (B)			25–50		
Molybdenum ppm (Mo)	0.04[D]–0.13[D]	0.14–0.30	0.45–1.00		

[D] Showing deficiency symptoms.
[T] Showing toxicity symptoms.

Chickpea
(Cicer arietinum)

Plant part Recently matured leaf
Growth stage Vegetative, i.e. pre-flowering

Nutrient	Deficient	Below normal	Normal	Above normal	Excess or toxic
Nitrogen % (N)	1.25^D–2.4^D	3.2–3.4	4.0–5.5	5.9?	
Phosphorus % (P)	0.10^D–0.13^D	0.24	0.29–0.55		
Potassium % (K)	1.0^D	1.6–1.8	2.0–3.6		
Calcium % (Ca)			1.3–2.2	2.8?	
Magnesium % (Mg)		0.15	0.35–0.65		
Sodium % (Na)			0.01–0.20		
Chloride % (Cl)			0.4–1.0	1.5	1.6–3.9^T
Copper ppm (Cu)			4–30		
Zinc ppm (Zn)			22–90		
Manganese ppm (Mn)			60–300	350?	
Boron ppm (B)			22–30		

D Showing deficiency symptoms.
T Showing toxicity symptoms.

Cotton
(Gossypium hirsutum)

Plant part   Uppermost mature leaf on vegetative stems
Growth stage   Pre-bloom to first bloom

Nutrient	Deficient	Below normal	Normal	Above normal	Excess or toxic
Nitrogen % (N)	2.3–2.5	3.0–3.3	3.5–4.7	5.0–6.0	
Phosphorus % (P)	0.13–0.15	0.21–0.23	0.25–0.50	0.55	0.91
Potassium % (K)	<1.2	1.2–1.3	1.5–3.0		
Calcium % (Ca)			2.2–3.8	4.5–5.0	
Magnesium % (Mg)		0.02–0.29	0.30–0.90	1.0–1.2	1.4[1]–1.5[1]
Sodium % (Na)			0.02–0.35	0.4–0.5	0.6
Chloride % (Cl)			0.5–1.5	1.7–2.0	2.2[T]–4.3[T]
Copper ppm (Cu)		3–4	5–30		
Zinc ppm (Zn)	7–13	16–20	25–60		
Manganese ppm (Mn)	8	15–20	25–500	1000–2000	4000
Boron ppm (B)	<16	16–19	20–100	120–250	

[1] Associated with low leaf potassium.
[T] Showing toxicity symptoms.

Cowpea
(Vigna unguiculata)

Plant part Recently matured leaf
Growth stage Pre-flowering

Nutrient	Deficient	Below normal	Normal	Above normal	Excess or toxic
Nitrogen % (N)	2.0–2.4	2.6–3.2	3.6–4.6		
Phosphorus % (P)	0.14[D]–0.20	0.22	0.28–0.48		
Potassium % (K)	0.5[D]–0.9[D]	1.1–1.2	1.7–3.0		
Calcium % (Ca)			1.5–3.0		
Magnesium % (Mg)			0.35–1.0	1.30[1]	
Sodium % (Na)			0.01–0.03		
Chloride % (Cl)			0.7–1.6	1.9	
Copper ppm (Cu)			8–20		
Zinc ppm (Zn)			22–70		
Manganese ppm (Mn)			70–300	350–600	

[1] Imbalance associated with low leaf potassium.
[D] Showing deficiency symptoms.

Desmodium (silverleaf and greenleaf)
(*Desmodium uncinatum* and *D. intortum*)

Plant part   Stripped leaves
Growth stage   Pre-flowering

Nutrient	Deficient	Below normal	Normal	Above normal	Excess or toxic
Nitrogen % (N)	2.3[D]	2.7	3.5–5.0		
Phosphorus % (P)	0.16[D]	0.20–0.24	0.25–0.40		
Potassium % (K)	0.36[D]	0.5–0.6	0.8–2.0		
Sulphur % (S)		0.13–0.18	0.20–0.40		
Calcium % (Ca)			1.0–2.0		
Magnesium % (Mg)			0.25–0.60		
Sodium % (Na)			<0.05		
Chloride % (Cl)			0.5–1.0	2.0	2.3[T]–4.2[T]
Copper ppm (Cu)		4–5	8–60		
Zinc ppm (Zn)		<20	20–70		
Manganese ppm (Mn)			40–350	675	

[D] Showing deficiency symptoms.
[T] Showing toxicity symptoms.

Dolichos lablab

Plant part   Young mature leaves
Growth stage   Pre-flowering

Nutrient	Deficient	Below normal	Normal	Above normal	Excess or toxic
Nitrogen % (N)	1.9	2.5–3.0	3.5–5.3		
Phosphorus % (P)		0.20–0.23	0.25–0.40	0.50	
Potassium % (K)	0.4[D]–0.8[D]	0.9	1.0–1.8		
Sulphur % (S)			0.21–0.35		
Calcium % (Ca)			1.2–1.7		
Magnesium % (Mg)			0.24–0.55	0.65	0.70[1]
Sodium % (Na)			<0.20		
Chloride % (Cl)			<1.0		
Copper ppm (Cu)			8–20		
Zinc ppm (Zn)			30–70		
Manganese ppm (Mn)			50–200		
Boron ppm (B)			30–50		
Molybdenum ppm (Mo)		0.14	0.3–1.0		

[1] Imbalance associated with potassium-deficient crop.
[D] Showing deficiency symptoms.

Faba beans
(Vicia faba)

Plant part   Recently matured leaf
Growth stage   Early flowering

Nutrient	Deficient	Below normal	Normal	Above normal	Excess or toxic
Nitrogen % (N)	1.6[D]–2.5[D]	3.6	4.3–5.0		
Phosphorus % (P)	0.16[D]	0.19–0.24	0.30–0.55	0.82	
Potassium % (K)	1.7	1.8–2.0	2.2–4.0		
Calcium % (Ca)			0.6–1.2		
Magnesium % (Mg)			0.24–0.50		
Sodium % (Na)			0.02–0.40	0.7	2.2[T]
Chloride % (Cl)			<0.8	0.9–1.2	
Copper ppm (Cu)			5–25		
Zinc ppm (Zn)			28–140		
Manganese ppm (Mn)			50–300	600–900	1000–2020[T]
Molybdenum ppm (Mo)		0.19	0.4–5.5		

[D] Showing deficiency symptoms.
[T] Showing toxicity symptoms.

Field pea
(Pisum sativum)

Plant part Youngest mature compound leaf
Growth stage Pre-flowering

Nutrient	Deficient	Below normal	Normal	Above normal	Excess or toxic
Nitrogen % (N)	1.4–2.8	3.2–3.4	4.0–5.5	5.8	
Phosphorus % (P)	0.18	0.24–0.27	0.30–0.55	0.65–0.85	
Potassium % (K)	<1.5	1.6–1.8	2.0–3.4		
Sulphur % (S)			0.20–0.40		
Calcium % (Ca)		0.7	0.9–2.0		
Magnesium % (Mg)	<0.16	0.16–0.20	0.20–0.55	>0.7	
Sodium % (Na)			<0.3	>0.4	
Chloride % (Cl)			<1.6	1.8–2.0	>2.0
Copper ppm (Cu)		4	6–25		
Zinc ppm (Zn)	<16	16–19	24–100	>100	
Manganese ppm (Mn)		15	30–400	500–800	>1000 (4000^T)
Boron ppm (B)		<20	25–60		
Molybdenum ppm (Mo)			0.3–1.0		

^{T} Showing toxicity symptoms.

Linseed
(Linum usitatissimum)

Plant part   Upper fully expanded leaves
Growth stage   Immediately pre-flowering

Nutrient	Deficient	Below normal	Normal	Above normal	Excess or toxic
Nitrogen % (N)	2.2	2.6–2.8	3.2–4.5		
Phosphorus % (P)	0.13–0.16	0.20–0.22	0.25–0.45		
Potassium % (K)	0.8	1.6–1.8	2.2–3.0		
Calcium % (Ca)		0.8	1.0–1.6	2.2	
Magnesium % (Mg)			0.30–0.65	0.95	
Sodium % (Na)			0.2–0.9	1.1–1.2	
Chloride % (Cl)			0.3–1.2		
Copper[1] ppm (Cu)		4	6–25	90	
Zinc[1] ppm (Zn)	9–14	15–18	20–90		
Manganese[1] ppm (Mn)			60–800		
Boron ppm (B)	15	19	25–60		

[1] Values for copper, zinc or manganese in leaves sprayed with fungicides or nutrient sprays containing trace elements cannot give a reliable guide to nutritional status even in washed leaves.

Lupins

(Lupinus angustifolius, L. albus and *L. luteus)*

Plant part   Youngest fully expanded leaf
Growth stage   Pre-flowering

Nutrient	Deficient	Below normal	Normal	Above normal	Excess or toxic
Nitrogen % (N)	1.1[D]–2.3[D]	2.5–2.8	3.2–6.0	7.0	
Phosphorus % (P)	0.12[D]–0.16	0.18–0.19	0.24–0.50		
Potassium % (K)	<1.1	1.1–1.4	1.6–3.5		
Calcium % (Ca)			1.0–3.0		
Magnesium % (Mg)		0.17	0.24–1.10		
Sodium % (Na)			0.02–0.35	0.4–0.7	1.0[T]–1.6[T]
Chloride % (Cl)			0.1–1.2		
Copper[1] ppm (Cu)			5–25		
Zinc[1] ppm (Zn)		10–15	25–100	300	
Manganese[1] ppm (Mn)			50–1200	1400–1600	1900[T]–16 000[T]
Boron ppm (B)	12	15–18	20–60		
Molybdenum ppm (Mo)			0.3–5.0	9.0	

[1] Values for copper, zinc or manganese in leaves sprayed with fungicides or nutrient sprays containing trace elements cannot give a reliable guide to nutritional status even in washed leaves.
[D] Showing deficiency symptoms.
[T] Showing toxicity symptoms.

Peanut
(Aracis hypogaea)

Plant part   Upper mature leaf
Growth stage   Pre-flowering to early bloom

Nutrient	Deficient	Below normal	Normal	Above normal	Excess or toxic
Nitrogen % (N)	1.3–2.5	2.8–3.0	3.5–5.0		
Phosphorus % (P)	0.13–0.15	0.12–0.23	0.25–0.55	>0.6	
Potassium % (K)		1.0–1.3	1.6–3.0		
Sulphur % (S)			0.20–0.35		
Calcium % (Ca)			1.2–2.4		
Magnesium % (Mg)			0.30–0.80		
Sodium % (Na)			0.01–0.10	>0.3	
Chloride % (Cl)			0.3–1.2	>1.5	
Copper ppm (Cu)		3	6–30		
Zinc ppm (Zn)		18–20	25–80	>80	
Manganese ppm (Mn)		25	50–300	350–700	>700
Boron ppm (B)		18–20	25–60		
Molybdenum ppm (Mo)		0.05	0.10–5.0		

Perennial ryegrass[1]
(Lolium perenne)

Plant part   Randomly plucked young mature leaves
Growth stage   Vegetative, during active growth

Nutrient	Deficient	Below normal	Normal	Above normal	Excess or toxic
Nitrogen % (N)			3.5–4.5	5.0–6.0	
Phosphorus % (P)	0.13	0.18–0.21	0.25–0.55	0.60–0.75	0.80?
Potassium % (K)	0.8–1.3	1.4–1.6	2.0–4.8	7.0	
Sulphur % (S)		0.18–0.22	0.25–0.45	0.55–0.65	
Calcium % (Ca)			0.20–0.60		
Magnesium % (Mg)	0.07–0.13	0.14	0.20–0.40		
Sodium % (Na)			0.1–0.5	0.7^2–1.5^2	1.8^2–2.3^2
Chloride % (Cl)			<2.0	2.0–2.6	3.6
Copper ppm (Cu)		4–5	6–15		
Zinc ppm (Zn)		10–13	15–70		
Manganese ppm (Mn)		16	30–400	650–1290	
Boron ppm (B)		5	10–20		
Molybdenum ppm (Mo)		0.11	0.25–3.7	8.1	

[1] Diagnosis of poor pasture growth is normally best made on the dominant legume species.
[2] Most high sodium values were associated with potassium deficiency (leaf K values of 0.8–1.6% K) rather than salinity.

Phalaris[1]
(Phalaris aquatica)

Plant part   Randomly plucked young mature leaves
Growth stage   Vegetative, during active growth 3–5 weeks of regrowth

Nutrient	Deficient	Below normal	Normal	Above normal	Excess or toxic
Nitrogen % (N)	<2.0	2.0–2.9	3.5–5.0		
Phosphorus % (P)	0.11–0.17	0.20–0.24	0.27–0.45		
Potassium % (K)	1.1–1.4	1.7–2.0	2.0–3.7		
Sulphur % (S)	<0.20	0.20–0.22	0.25–0.50		
Calcium % (Ca)			0.14–0.40		
Magnesium % (Mg)		0.16	0.22–0.35		
Sodium % (Na)			0.1–0.5	0.7–0.8	1.0–1.3
Chloride % (Cl)			<2.0	2.5–2.8	
Copper ppm (Cu)		1–2	4–12		
Zinc ppm (Zn)		11–14	15–80		
Manganese ppm (Mn)			30–300	490	
Boron ppm (B)			8–15		
Molybdenum ppm (Mo)			0.15–1.6		

[1] Diagnosis of poor pasture growth is normally best made on the dominant legume species.

Tobacco
(Nicotiana tabacum)

Plant part   Most recently matured leaf
Growth stage   Pre-flowering, 40–80 days after emergence

Nutrient	Deficient	Below normal	Normal	Above normal	Excess or toxic
Nitrogen % (N)		3.5	4.0–5.4		
Phosphorus % (P)		0.20–0.22	0.25–0.50		
Potassium % (K)		2.0	2.5–5.0	5.8–6.8	
Calcium % (Ca)			1.3–2.3	2.9–3.5	
Magnesium % (Mg)	0.11	0.18	0.25–0.90	1.2–1.4	
Sodium % (Na)			0.01–0.15	0.4–0.6	0.8[T]
Chloride % (Cl)			<2.5	3.0–3.5	>3.5[1]
Copper ppm (Cu)		2–3	4–55		
Zinc ppm (Zn)			20–70		
Manganese ppm (Mn)			35–350	450–850	1290[T]–1420[T]
Boron ppm (B)		15–20	25–55		
Molybdenum ppm (Mo)		<0.4	0.4–3.7		

[1] Quality can suffer when leaf chloride exceeds this value.
[T] Sample from crop showing toxicity symptoms.

Vetch
(*Vicia* spp.)

Plant part   Upper tops
Growth stage   Pre-flowering

Nutrient	Deficient	Below normal	Normal	Above normal	Excess or toxic
Nitrogen % (N)	1.2[D]–2.4[D]	3.0–3.4	4.4–6.0		
Phosphorus % (P)		0.17–0.19	0.30–0.55	0.7–0.8	
Potassium % (K)	<1.0[D]–1.6	2.5–2.8	3.0–4.2		
Calcium % (Ca)			0.6–1.8		
Magnesium % (Mg)		0.16	0.25–0.50	0.7	
Sodium % (Na)			0.01–0.20	0.4	
Chloride % (Cl)			0.1–1.0	1.4	1.7
Copper ppm (Cu)			8–20		
Zinc ppm (Zn)			30–100	250	
Manganese ppm (Mn)			50–350	500	
Boron ppm (B)	8[D]		25–30		
Molybdenum ppm (Mo)			0.5–3.0		

[D] Showing deficiency symptoms.